Pioneers in Human Anatomy

About this Book

Instructors and students reviewing, adopting, or adapting this textbook are encouraged to provide feedback (or report errors) online at: https://bit.ly/feedback-applied-human-neuroanatomy.

Pioneers in Human Anatomy

Gross Anatomy, Neuroanatomy, Microscopic Anatomy, Developmental Anatomy

Michael F. Nolan, Ph.D., P.T.
John P. McNamara, M.S., D.C.

Virginia Tech Carilion School of Medicine
in association with Virginia Tech Publishing Blacksburg, VA

A free version of this book can be downloaded from
https://doi.org/10.21061/pioneers-in-human-anatomy

Front and back cover images courtesy of Wellcome Library, London.

First published in 2023 by the Virginia Tech Carilion School of Medicine in association with Virginia Tech Publishing

Virginia Tech Carilion School of Medicine
2 Riverside Circle
Roanoke, VA 24016 USA

Virginia Tech Publishing
University Libraries at Virginia Tech
560 Drillfield Drive
Blacksburg, VA 24061 USA

Cataloging-in-Publication Data

Names: Nolan, Michael F., author | McNamara, John P., author.

Title: Pioneers in Human Anatomy / Michael F. Nolan and John P. McNamara.

Description: Roanoke, VA : Virginia Tech Carilion School of Medicine in association with Virginia Tech Publishing, 2023. | Summary: “This book provides an opportunity for those who may be interested in those pioneers in anatomy or who may have learned (or forgotten) some eponymonic descriptions in the past, to test their knowledge (or memory) of these historical designations or to link particular individuals to anatomical structures and features previously unknown.” – Introduction.

Identifiers: ISBN: 978-1-957213-67-5 (PDF) | ISBN: 978-1-957213-57-6 (pbk) |
DOI: https://doi.org/10.21061/ pioneers-of-human-anatomy

Subjects: HUMAN ANATOMY / Laboratory manuals. | HUMAN ANATOMY / Problems and Exercises.

Table of Contents

Introduction

One cannot study medicine without encountering the names of those individuals who first discovered and/or described a particular structure, procedure, disease, sign, symptom or instrument of diagnosis or treatment. It seems as though eponyms in medicine have always been around and it seems unlikely that they will ever be totally abandoned or eliminated. Eponymonic designations are a way of recognizing and giving credit for work sometimes spanning a lifetime.

Some eponyms have come into the literature because that individual was the first to report on a particular topic (*aqueduct of Sylvius*) or may have provided the most definitive description of the topic (*Parkinson disease*). Others recognize the committed effort or discovery of a single individual (*islets of Langerhan*) while others signify the combined or collaborative efforts of two or more dedicated persons (*Marchiafava-Bignami disease*). Some refer to normal structures (*Hunter's canal*) or functions (*Starling's law*), while many others relate to abnormal observations (*Babinski sign*). Countless numbers are associated with diseases (Tay-Sachs disease) or injuries (*Colle's fracture*), procedures for evaluating functions (*McMurray's test*) or techniques for treating specific disease or injuries (*Epply maneuver*). Eponymonic designations are used to identify tools and surgical instruments invented for particular purposes (*Kerrison Rongeur*) and surgical procedures (*Whipple procedure*). Some eponyms are associated with other, frequently more descriptive terms (*Poupart's ligament / inguinal ligament*) while others indicate structures, processes or procedures for which no other name exists (*McBurney's point*). As can be appreciated from this brief synopsis, eponyms are well ingrained into almost all facets of medicine.

In recent years some efforts have been made to replace eponyms in the anatomical literature with more descriptive terms; an understandable goal in the minds of some and an undesirable and objective in the minds of others. To those who may favor eliminating eponymonic designations we, of a more historical view and leaning, say "A pox upon you and your house – how dare you strip our professional lexicon of such an elegant and deserving way of acknowledging those upon whose shoulders we stand!" We believe that eponyms, like rock and roll, are here to stay.

So, in order to assist in stalling this outrageously unthinkable effort, we offer here for your enjoyment and enrichment an opportunity, in a quiz format, to link descriptions of various anatomical structures with the names of those pioneers in the anatomical sciences, some well-known and others less well-known, with whom they are associated. We decided to make this journey through the world of eponymonic descriptions more enjoyable (and even challenging) by presenting this material in a quiz format. We believe that quizzes are fun and the satisfaction of getting a few correct can even be rewarding.

This collection of free-answer quiz questions is divided into three sections. In the first section a brief description of an anatomical structure, feature, tissue, cell or pathway is presented followed by a blank line in which the reader may insert the name(s) of the individual(s) commonly associated with that structure, feature, tissue or pathway. The descriptions and individuals included here are limited to those associated with the four traditional sub-disciplines of anatomy: gross anatomy, neuroanatomy, microscopic anatomy and developmental anatomy. Within these major anatomical subdisciplines, the descriptions are further categorized to better organize the material. For example, in the chapter on Gross Anatomy, questions are further divided into those associated with different body regions. In the chapter on Microscopic Anatomy, descriptions are assembled in groups based on organ system. In the chapter on Neuroanatomy the descriptions include categories such as nuclei, ganglia, peripheral nervous system and tracts. We believe this organizational format will help the reader who might be interested only in descriptions and individuals associated with a particular body region or system.

In the second section the descriptions are repeated with the name(s) of the individuals associated with each particular structure included on the line provided with each description. This section serves as a type of answer key. For most descriptions only one individual is or has been associated with a particular structure. However, in some cases more than one individual has been recognized in connection with a particular anatomical entity. We make no effort to adjudicate any disputes regarding priority or originality in these matters, but rather list multiple names when they are found in commonly used and easily available resources and when we can find no reliable evidence to exclude them.

The third section includes an alphabetical listing of the individuals listed in section two. This list is unfortunately, but necessarily incomplete. Birth and death dates, particularly for those who lived in the 16th, 17th and 18th century may not have been included due to the absence of, or our inability to locate any reliable information on this head. We have found this listing useful in finding additional, often interesting information about these worthy individuals. Simply enter the individual's name into a search engine, press the enter key and like magic, a selection of sites appears that allows the reader to explore further the lives and work of these worthy individuals.

Many resources were consulted in undertaking this project. These include the published literature in which particular anatomical structures and their eponymonic designation are mentioned or discussed. Some material was obtained from textbooks, including a collection of very old textbooks published the middle and late 19th and early 20th centuries which we have been privileged to examine over the years and which have been a delight to use. Other contributions were obtained through internet searches. We acknowledge the fallibility of some of the resources we have used and accept full responsibility for any errors that might have been included. Additionally, we acknowledge that the spelling of the names of several of these individuals may vary in the literature, typically reflecting the origin and the language of the resources used in assembling this information. Every effort has been made to be as accurate as possible in this work. We ask your understanding if the spellings and completeness of the names and the birth and death dates included differ from what you might be familiar with. We would be pleased hear from readers on this matter and to correct any errors in subsequent editions.

This work in not intended to be encyclopedic or exhaustive in scope and as such is not to be considered as a reference resource. It is not intended to be used as a textbook or learning resource in an anatomy course or graduate program representing subject matter that students would be expected to memorize and use in their professional lives. Nor is it intended to serve as a library of questions that might be used on examinations to be used as a means of distinguishing among student populations. It is rather intended to provide an opportunity for those who may be interested in those pioneers in anatomy or who may have learned (or forgotten) some eponymonic descriptions in the past, to test their knowledge (or memory) of these historical designations or to link particular individuals to anatomical structures and features previously unknown. While students and teachers of anatomy may might have a professional interest in knowing who, in the history of anatomy, has been associated with a particular structure, others including medical historians, biographers and medical artist might also find this information of interest.

In deciding on the format to use for this material, we thought that the free answer quiz format might be both engaging and fun and after trying it out ourselves and with our colleagues, we found this to be correct; satisfying when we were able to link the description with an individual and sadly frustrating when we failed to do so. Being anatomists ourselves, we found this approach to have a certain appeal and we thought it ultimately a rather good idea.

We hope you enjoy attempting to link the many anatomical structures listed here with the individuals with whom they are forever connected. We hope also that you will use them from time to time as may seem appropriate and experience as we have, the pleasure and intellectual satisfaction of recognizing those who have labored before us.

MFN
JPM

Section One - Quizzes

Gross Anatomy - Head and Neck

Directions: Indicate the name of the individual(s) in column "B" that are associated with the structure/description provided in column "A."

	A	B
1.	maxillary sinus	________________
2.	facial canal	________________
3.	lesser (small) sublingual glands	________________
4.	posterolateral pharyngeal recess	________________
5.	pterygoid canal and its vessels	________________
6.	pharyngeal lymphatic ring	________________
7.	submandibular duct	________________
8.	cuneiform cartilage of larynx	________________
9.	opening into cavernous sinus for passage of VI nerve	________________
10.	smooth muscle of eyelids	________________
11.	parotid duct	________________
12.	pharyngo-tympanic tube	________________
13.	tarsal glands of the eyelid	________________
14.	fascial sheath of the eye	________________
15.	laryngeal ventricle	________________
16.	aryepiglottic muscle	________________
17.	superior nasal concha	________________
18.	accessory lacrimal glands	________________
19.	petroclinoid ligament	________________
20.	carotid tubercle	________________
21.	vomeronasal cartilage	________________
22.	corniculate cartilage of the larynx	________________

23. space between dural folds that houses trigeminal ganglion ____________________
24. temporo-squamousal suture ____________________
25. suspensory ligament of the lens ____________________
26. cleft in meningeal dura housing trigeminal ganglion ____________________
27. cartilage of the 2nd branchial arch ____________________
28. small bones found in sutures of the skull ____________________
29. canal in temporal bone for auricular branch of X ____________________
30. columns in the osseous spiral lamina of the cochlea ____________________
31. foramen cecum of tongue ____________________
32. space between the upper border of levator veli palatini and base of the skull ____________________
33. tympanic notch on tympanic part of temporal bone ____________________
34. palpebral portion of the lacrimal gland ____________________
35. incisive foramina of the hard palate ____________________
36. venous sinus of the sclera ____________________
37. lacrimal fold ____________________
38. ductus reuniens ____________________
39. thickened lower part of the bulbar sheath ____________________
40. fold on posterior pharyngeal wall ____________________
41. tensor tarsi (lateral part of orbicularis oculi) ____________________
42. small swelling on posterior helix of ear ____________________
43. lowest thyropharyngeus fibres above the cricopharyngeus muscle ____________________
44. five artery anastomoses on anterior nasal septum ____________________
45. line from inferior margin of orbit to external auditory canal ____________________
46. vein connecting pterygoid plexus and cavernous sinus ____________________
47 membrane fold in lacrimal duct near punctum ____________________
48. retropharyngeal lymph nodes ____________________

49. tympanic canal ____________________
50. sublingual gland duct ____________________
51. sphenoid conchae ____________________
52. space above jugular notch ____________________
53. small defect on posterior edge of tympanic ring ____________________
54. tubular cavity in nasal septum ____________________
55. accessory lacrimal glands ____________________
56. deep lingual arterial anastomosis ____________________
57. space in iridocorneal angle ____________________
58. thyroglossal duct ____________________
59. buccal fat pad ____________________
60. anterior process of malleus ____________________
61. helicotrema ____________________
62. opening in osseous part of external auditory meatus ____________________
63. minor sublingual glands ____________________
64. petrotympanic fissure ____________________
65. scaphoid fossa in sphenoid bone ____________________
66. ciliary muscle ____________________
67. membrane lining the maxillary sinus ____________________
68. alar ligament ____________________
69. parietal emissary veins ____________________
70. lacrimal part of orbicularis oculi ____________________
71. triangle involving hyoglossus, digastric and hyoid bone ____________________
72. pharyngeal ridge with contraction of superior constrictor ____________________
73. pyramidal lobe of thyroid gland ____________________
74. sublingual bursa ____________________
75. muscle fibers in check ligament of the eye ____________________
76. tentorial notch ____________________
77. central artery of the retina ____________________

78. lateral capsular ligament of temporomandibular joint ____________

79. hiatus for the lesser petrosal nerve ____________

80. thyroid ima artery ____________

81. auricular ligaments ____________

81. hiatus for the greater petrosal nerve ____________

82. anterior lingual gland ____________

83. laryngeal ventricle ____________

84. meridional fibers of the ciliary muscle ____________

85. suspensory ligament of the thyroid gland ____________

86. tubal tonsil ____________

87. cartilage attached to vocal process of arytenoid ____________

88. risorius muscle ____________

89. line passing between opisthion and basion ____________

90. sphenoidal concha ____________

91. depression in lacrimal sac for lacrimal duct ____________

92. subarachnoid space around optic nerve ____________

93. petro-occipital fissure ____________

94. parotid duct ____________

95. optic disc ____________

96. scalenus minimus ____________

97. spiral canal of cochlea ____________

98. diploic veins ____________

99. olfactory glands ____________

100. innermost layer of the choroid of the eye ____________

101. ligament between pars tensa and pars flaccida ____________

102. small recess in middle ear cavity ____________

103. true vocal folds ____________

104. retro-lingual space ____________

105. gap in petrotympanic fissure for chorda tympani ____________

106. highest nasal concha ____________

107. ligament of the malleus ____________

108. superficial portion of orbicularis oculi ____________________

109. small mucous gland in floor of the mouth ____________________

110. fovea centralis ____________________

111. clivus ____________________

112. pretarsal segment of orbicularis oculi ____________________

113. endolymphatic sac ____________________

114. lateral tubercle of the orbit ____________________

115. mastoid air sinuses ____________________

116. anterior part of fascial sheath of the eye ____________________

117. suspensory ligament of the esophagus ____________________

118. parathyroid gland ____________________

119. mastoid fontanelle ____________________

120. pterygopalatine fossa ____________________

121. small cavity on floor of external auditory canal ____________________

122. small cartilage on anterior fold of vocal cord ____________________

123. bursa posterior to hyoid bone ____________________

124. superior transverse ligament of the eye ____________________

125. potential space between ciliary zonule and vitreous body ____________________

126. anterior limiting lamina of cornea ____________________

127. notches on cartilage of external auditory meatus ____________________

128. conjunctival glands ____________________

129. sphenoparietal sinus ____________________

130. pterygospinous ligament and process ____________________

131. carotid space (triangle) ____________________

132. cochlear duct ____________________

133. pharyngeal tonsil ____________________

134. small orbital muscle that pulls eye forward ____________________

135. short ciliary arteries on sclera near optic nerve ____________________

136. anterior tympanic artery ____________________

137. taste bud ____________________

138. pars flaccida ________________

139. horizontal cartilage near nasal septum ________________

140. cutaneous muscle of the lip ________________

142. endolymphatic recess ________________

143. space containing carotid bifurcation ________________

144. tonsillar fossa ________________

145. episcleral space ________________

146. connective tissue fibers bounding pars flaccida ________________

147. triangle involving digastric, mylohyoid and hypoglossal nerve ________________

148. median atlanto-occipiral ligament ________________

149. tympanomeningeal fissure ________________

150. ansa subclavian ________________

151. epithelial tissue near angle of mandible by buccal nerve ________________

152. bulb of the jugular vein ________________

153. canal in vitreus from optic disc to lens ________________

154. small venous plexus between canal of Schlemm and space of Fontana ________________

155. space between tongue and sublingual gland ________________

156. exocrine salivary glands ________________

157. arterial plexus of the nasal septum ________________

158. suprameatal triangle / mastoid fossa ________________

159. left supraclavicular lymph node ________________

160. triangular area on posterior proximal esophagus ________________

Gross Anatomy - Limbs

Directions: Indicate the name of the individual(s) in column "B" that are associated with the structure/description provided in column "A".

	A	B
1.	iliofemoral ("Y") ligament	____________________
2.	adductor canal	____________________
3.	dorsal tubercle of the radius	____________________
4.	inguinal ligament	____________________
5.	posterior meniscofemoral ligament	____________________
6.	supinator arch	____________________
7.	fibula-talo-calcaneal ligament	____________________
8.	iliopectineal fascia	____________________
9.	line joining ASIS to ischial tuberosity	____________________
10.	common tendon of gastrocnemius and soleus	____________________
11.	oblique popliteal ligament	____________________
12.	transverse tarsal joint	____________________
13.	palmar fascia of the hand	____________________
14.	superior glenohumeral ligament	____________________
15.	suspensory ligament of the axilla	____________________
16.	anterior meniscofemoral ligament	____________________
17.	tarsometatarsal joint	____________________
18.	iliotibial tract (band)	____________________
19.	femoral triangle	____________________
20.	oblique cord of proximal radio-ulnar joint	____________________
21.	abnormal abductor digiti minimi of the hand	____________________
22.	tunnel transmitting the ulnar nerve at the wrist	____________________
23.	ligament forming tunnel through which passes the median nerve and brachial artery	____________________
24.	cutaneous ligaments of the digits	____________________

25. transverse band of the ulnar collateral ligament ________________

26. canals within bones containing vessels ________________

27. lacunae within bones containing osteoclasts ________________

28. inguinal lymphatic nodes adjacent to femoral canal ________________

29. lateral tubercle of tibia for attachment of the IT band ________________

30. angle of inclination of the femur ________________

31. posterior process of the talus (astragalus) ________________

32. lymph node at the femoral ring ________________

33. inferior calcaneal nerve ________________

34. posterior process of the talus ________________

35. fibrous roof over cubital tunnel ________________

36. lower fibers of posterior sacroiliac joint ________________

37. articularis genu muscle ________________

38. proximal of the two palmar limb creases ________________

39. coracobrachialis muscle ________________

40. semimembranosus bursa ________________

41. ligament from posterior superior iliac spine to
2nd transverse tubercle on sacrum ________________

42. calcar femorale ________________

43. lacertus fibrosus ________________

44. articular cartilage of the acromioclavicular joint ________________

45. dorsal talonavicular ligament ________________

46. tuberosity of the 5th metatarsal ________________

47. fibular collateral ligament ________________

48. septal tissue in femoral canal ________________

49. supraclavicular lymph nodes on left side ________________

50. superficial transverse ligament of the palm ________________

51. accessory anterior inferior talofibular ligament ________________

52. plantar ligament of the foot ________________

53. line passing through great toe and mid calcaneus ________________

54. split in flexor digitorum superficialis tendon for tendon of flexor digitorum profundus ____________________

55. longitudinal fibers of articular capsule of hip joint ____________________

56. volar carpal ligament ____________________

57. space between pronator quadratus and FDP in forearm ____________________

58. location of posterior tibial artery behind medial malleolus ____________________

Gross Anatomy - Thorax

Directions: Indicate the name of the individual(s) in column "B" that are associated with the structure/description provided in column "A".

	A	B
1.	suspensory ligament of the breast	____________
2.	sternal angle	____________
3.	sebaceous glands of the areola	____________
4.	conduction fibers of the heart	____________
5.	suprapleural membrane (fascia)	____________
6.	valve of the coronary sinus	____________
7.	atrio-ventricular bundle	____________
8.	attachment for anterior scalene on the first rib	____________
9.	long thoracic nerve	____________
10.	foramen ovale of the heart	____________
11.	ductus (ligamentum) arteriosum	____________
12.	space above the jugular notch	____________
13.	sino-atrial node	____________
14.	oblique vein of the left atrium	____________
15.	smooth muscle fibers of the smallest bronchi	____________
16.	atrio-ventricular node	____________
17.	retromammary space	____________
18.	valve of the inferior vena cava	____________
19.	valve of the coronary sinus	____________
20.	fold in left superior vena cava in pericardial sac	____________
21.	third intercostal space on the left sternal border (S2)	____________
22.	lymphatics in areola of nipple	____________
23.	sternopericardial ligament	____________
24.	partial duplication of thoracic duct in thorax	____________
25.	nodules on edge of semilunar valves	____________
26.	interspinous plane	____________

27. thoracic duct ____________________

28. transverse pericardial sinus ____________________

29. triangular area in wall of right atrium ____________________

30. pulmonary ridge ____________________

31. lacteal veins ____________________

32. innominate cardiac vein ____________________

33. mammary tissue extending into axilla ____________________

34. anastomosis between circumflex artery and right coronary artery ____________________

35. largest openings of the small cardiac veins ____________________

36. pulmonary circulation ____________________

37. space between costal and sternal attachments of diaphragm ____________________

38. intervenous tubercle of right ventricle ____________________

39. infraclavicular fossa ____________________

40. small air-filled sacs on surface of lung ____________________

41. arterial conduction system of the heart ____________________

42. interpectoral lymph nodes ____________________

Gross Anatomy - Abdomen

Directions: Indicate the name of the individual(s) in column "B" that are associated with the structure/description provided in column "A".

	A	B
1.	main pancreatic duct	______________
2.	accessory pancreatic duct	______________
3.	hepatorenal recess	______________
4.	epiploic foramen	______________
5.	lacunar ligament	______________
6.	anterior renal capsule (fascia)	______________
7.	spiral fold of cystic duct	______________
8.	inguinal triangle	______________
9.	hepatopancreatic sphincter	______________
10.	lumbar triangle	______________
11.	lymphatic nodules of the ileum	______________
12.	suspensory ligament of the duodenum	______________
13.	hepatopancreatic ampulla	______________
14.	fatty layer of superficial abdominal fascia	______________
15.	fibrous layer of superficial abdominal fascia	______________
16.	ileal diverticulum	______________
17.	insulin secreting cells of pancreas	______________
18.	iliocecal fold	______________
19.	capsule of the liver	______________
20.	caudate lobe of the liver	______________
21.	cisternal chyli	______________
22.	supraduodenal artery	______________
23.	mucosal membrane folds in the ileum	______________
24.	vein crossing pyloduodenal junction	______________
25.	sternocostal triangle (foramen)	______________
26.	lumbocostal trigone (foramen)	______________

27. posterior renal capsule (fascia) ________________
28. transpyloric plane ________________
29. pudendal canal ________________
30. iliofemoral ligament ________________
31. capsule surrounding the glomerulus of the kidney ________________
32. glands secreting digestive enzymes in duodenum ________________
33. upper part of the pectineal fascia ________________
34. dilatation at the neck of the gall-bladder ________________
35. "u" shaped region of the nephron ________________
36. circular mucous folds of the small intestine ________________
37. inguinal ligament ________________
38. ileocecal fold ________________
39. ileocecal valve ________________
40. reflected part of the inguinal ligament ________________
41. linea semilunaris ________________
42. spiral valves of the cystic duct ________________
43. tubular structures in the wall of the gallbladder ________________
44. conjoint tendon of the transversus abdominis muscle ________________
45. small tubular structures in the wall of the gallbladder ________________
46. renal collecting ducts ________________
47. perirenal fascia ________________
48. retro-aortic left renal vein ________________
49. cysto-hepatic triangle ________________
50. small veins of the falciform ligament ________________
51. semilunar line of the abdominal wall ________________
52. line connecting gall bladder and IVC on ventral liver ________________
53. uncovertebral joints ________________
54. surface location for inflamed appendix ________________
55. ischioanal fossa ________________
56. ischiorectal fossa ________________

57. area bounded by lower border L lung, ant border of spleen, L costal margin, inf margin L lobe of liver ____________
58. superior hypogastric plexus ____________
59. small cutaneous veins around umbilicus ____________
60. ileal sphincter ____________
61. central anastomotic mesenteric artery ____________
62. middle lobe of prostate gland ____________
63. mucous membrane at base of appendix ____________
64. celiac trunk ____________
65. arterial anastomosis between SMA and IMA ____________
66. fissure on liver between right lobe and caudate lobe ____________
67. esophagogastric angle ____________
68. paraaortic bodies ____________
69. deep inguinal lymph nodes ____________
70. central veins of the liver ____________
71. small mucous glands of the biliary mucosa ____________
72. uncinated process of the pancreas ____________
73. frenulum of the ileal orifice ____________
74. intestinal glands ____________
75. capsule of the spleen ____________
76. inguinal canal ____________
77. interfoveolar ligament ____________
78. tongue of right lobe of liver extending downward ____________
79. circular membranous folds of small intestine ____________
80. superior lumbar triangle ____________
81. precolic fascia ____________
82. iliopubic ligament ____________
83. retroperitoneal veins ____________
84. falciform margin of saphenous opening ____________
85. line between anterior superior iliac spine and umbilicus ____________

86. mesenterico-parietal fascia ____________________
87. transversalis fascia ____________________
88. anterior layer of rectus sheath below arcuate line ____________________
89. inferior duodenal fossa/recess ____________________
90. superior mesenteric artery ____________________
91. esophagogastric angle ____________________
92. intrahepatic bile ductules ____________________
93. sling-like collar at gastroesophageal junction ____________________

Gross Anatomy - Pelvis and Perineum

Directions: Indicate the name of the individual(s) in column "B" that are associated with the structure/description provided in column "A".

	A	B
1.	anal columns	____________
2.	perineal fascia	____________
3.	bulbourethral glands	____________
4.	pudendal canal	____________
5.	deep fascia of the penis	____________
6.	rectouterine (rectovesical) pouch	____________
7.	transverse rectal folds	____________
8.	urethral glands	____________
9.	persistant processus vaginalis in females	____________
10.	retro-pubic space	____________
11.	paraurethral glands	____________
12.	greater vestibular glands	____________
13.	fascia separating prostate from rectum	____________
14.	sphincter of the urethra	____________
15.	cardinal ligament of the uterus	____________
16.	inter-ureteric ridge	____________
17.	trigone of the urinary bladder	____________
18.	rectosacral fascia	____________
19.	rectal valves	____________
20.	uterine tubes	____________
21.	perineal body	____________
22.	anococcygeal body	____________
23.	iliopectineal fascia	____________
24.	ampulla of the uterine tube	____________
25.	space in the bulbous portion of the urethra	____________
26.	middle lobe of the prostate gland	____________

27. openings in the mucous glands of the male urethrae ____________

28. circular muscle in the fundus uterus ____________

29. retro-renal fascia ____________

30. watershead between anterior and posterior branches of the renal arteries ____________

31. deep perineal fascia ____________

32. lateral edge of trigone area of urinary bladder ____________

33. coccygeal body ____________

34. nerve ganglion in lateral wall of cervix ____________

36. remnant of peritoneal processus vaginalis ____________

37. cremaster muscle ____________

38. mucous secreting glands lateral to lower vagina ____________

39. superficial fascia of the perineum ____________

40. lobules of epididymis ____________

41. subperitoneal fascia anterior to external iliac A. ____________

42. transverse line marking sacral promontory ____________

43. perineal membrane ____________

44. puboprostatic ligament ____________

45. navicular fossa of urethra ____________

45. longitudinal muscle of rectal wall ____________

46. deep fibers of dartos muscle in scrotal septum ____________

47. ovarian fossa ____________

48. perianal lymphatic plexus ____________

49. prostatic venous plexus ____________

50. cremasteric fascia ____________

51. rectovaginal septum ____________

52. round ligament of the uterus ____________

53. ischioanal fossa ____________

54. connective tissue fat in labia majora ____________

55. mesentery from broad ligament to appendix ____________

56. fundiform ligament of the clitoris ____________

57. paraurethral ducts ______________

58. scrotal raphe ______________

59. valve of the navicular fossa ______________

60. middle transverse rectal fold ______________

61. uvula of the bladder ______________

62. posterior urethral valve ______________

63. suspensory ligament of the ovary ______________

64. space between urinary bladder and intramural part of ureter ______________

65. smegma secreting glands of penis ______________

66. superficial perineal space ______________

67. gubernaculum testes ______________

68. lower 1/3 of the anal canal ______________

69. frenulum of ileocecal valve ______________

70. transverse perineal ligament ______________

71. isthmus of the ureter ______________

72. CT between bladder and anterior vaginal wall ______________

73. folds of peritoneum containing rectouterine muscle ______________

Neuroanatomy - Brain and Spinal Cord

Directions: Indicate the name of the individual(s) in column "B" that are associated with the structure/description provided in column "A".

	A	B
1.	cerebral aqueduct	____________
2.	lateral aperture of the IV ventricle	____________
3.	midline aperture of the IV ventricle	____________
4.	reticular process of the cord (lateral horn)	____________
5.	ventriculus terminalis (of the cord)	____________
6.	foramen cecum (pons/medullary junction)	____________
7.	tuberculum cinereum	____________
8.	arachnoid from dorsum sellae to mammillary body	____________
9.	interpeduncular fossa	____________
10.	small depression just anterior to obex	____________
11.	oligodendrocyte/Schwann cell border	____________
12.	hippocampus	____________
13.	interventricular foramen	____________
14.	cerebellar granular layer	____________
15.	superior medullary velum	____________
16.	cavum septum pellucidum	____________
17.	pons	____________
18.	microglia	____________
19.	lamina I of spinal cord	____________
20.	substantia gelatinous	____________
21.	tapetum	____________
22.	anterior part of the III ventricle	____________
23.	superior anastomotic vein	____________
24.	inferior anastomotic vein	____________
25.	great cerebral vein	____________
26.	confluence of sinuses	____________

27. arterial circle at base of brain ____________________
28. large spinal artery at T9 –T12 level ____________________
29. cortical areas designated by number ____________________
30. expressive language area ____________________
31. receptive language area ____________________
32. substantia innominate ____________________
33. laminae of white fibers in cerebral cortex ____________________
34. stria of the fourth ventricle ____________________
35. quadrigeminal cistern ____________________
36. paraolfactory area ____________________
37. olfactory islands ____________________
38. tail of the dentate gyrus ____________________
39. pontocerebellar cistern ____________________
40. reticular formation ____________________
41. olfactory bulb ____________________
42. insular lobe ____________________
43. trigone of the lateral lemniscus ____________________
44. tuberculum cinereum ____________________
45. cerebral basal vein ____________________
46. cistern of the lateral fissure ____________________
47. inferior medullary velum ____________________
48. centrum semiovale ____________________
49. artery of the pterygoid canal ____________________
50. cavity of the septum pellucidum ____________________
51. subarachnoid space around cerebral vessels ____________________
52. intercavernous (circular) sinus ____________________
53. arachnoid granulations ____________________
54. terminal ventricle of the spinal cord ____________________
55. calamus scriptorus ____________________
56. linea splendens ____________________
57. single artery from PCA supplying thalamus and

midbrain bilaterally ____________________

58. band of the dentate gyrus ____________________

59. substantia reticularis alba ____________________

60. cornu ammonis ____________________

61. arbor vitae (of cerebellum) ____________________

Neuroanatomy - Sulci and Gyri

Directions: Indicate the name of the individual(s) in column "B" that are associated with the structure/description provided in column "A".

	A	B
1.	central sulcus	____________
2.	horizontal fissure	____________
3.	superior transverse temporal gyrus	____________
4.	left middle frontal gyrus	____________
5.	rhinal sulcus	____________
6.	hypothalamic sulcus	____________
7.	gyrus descendens (lateral occipital lobe)	____________
8.	circular sulcus	____________
9.	supraorbital sulcus	____________
10.	incisura temporalis	____________
11.	gyrus olfactorio-orbitalis	____________
12.	subcallosal gyrus	____________
13.	dentate gyrus	____________
14.	supracallosal gyrus (induseum griseum)	____________
15.	sulcus intermedius secundus	____________
16.	intraparietal sulcus	____________
17.	insular cortex	____________
18.	gyrus lateral to fascia dentate	____________
19.	fissure below splenium of corpus callosum	____________
20.	occipital angle of the parietal lobe	____________
21.	medial olfactory gyrus	____________
22.	paraolfactory area	____________
23.	hippocampus	____________
24.	transverse gyrus of the insula	____________
25.	vertical fissure posterior to angular gyrus	____________
26.	inferior frontal gyrus	____________

27. sulcus from lower central sulcus to occipital lobe ____________________
28. subsplenial gyrus ____________________
29. diagonal band ____________________
30. gyrus fornicatus ____________________
31. transverse parietal sulcus ____________________
32. first intermediate sulcus (parietal lobe) ____________________
33. second intermediate sulcus ____________________
34. anterior transverse temporal sulcus ____________________
35. cingulate sulcus ____________________
36. parietooccipital sulcus ____________________
37. preoccipital notch ____________________
38. diagonal sulcus (frontal lobe) ____________________
39. radiate sulcus (frontal lobe) ____________________
40. frontomarginal sulcus ____________________
41. transverse insular sulcus ____________________
42. parietooccipital arc ____________________
43. pars triangularis (frontal lobe) ____________________
44. pars opercularis ____________________
45. superior temporal gyrus (posterior part) ____________________

Neuroanatomy - Ganglia

Directions: Indicate the name of the individual(s) in column "B" that are associated with the structure/description provided in column "A".

	A	B
1.	trigeminal ganglion	________
2.	pterygopalatine ganglion	________
3.	petrosal (inferior) ganglion of IX	________
4.	vestibular ganglion	________
5.	jugular (superior) ganglion of IX	________
6.	cells of the sympathetic ganglia	________
7.	otic ganglion	________
8.	spiral ganglion	________
9.	rudimentary ganglion cells on CN XII	________
10.	submandibular ganglion	________
11.	sympathetic cells near coronary sinus entry into atrium right atrium	________
12.	cardiac ganglia	________
13.	celiac ganglion	________
14.	myenteric ganglia of gut	________
15.	submucosal ganglia of gut	________
16.	cochlear ganglion	________
17.	carotid ganglion	________
18.	autonomic ganglia of the stomach	________
19.	ganglion impar	________
20.	ganglion of the superior alveolar nerve	________

Neuroanatomy - Nuclei

Directions: Indicate the name of the individual(s) in column "B" that are associated with the structure/description provided in column "A".

	A	B
1.	nucleus gracilis	______
2.	nucleus cuneatus	______
3.	midbrain parasympathetic nucleus	______
4	ventral posterior medial nucleus of thalamus	______
5.	lateral (external) cuneate nucleus	______
6.	interpeduncular nucleus	______
7.	superior vestibular nucleus	______
8.	medial vestibular nucleus	______
9.	lateral vestibular nucleus	______
10.	inferior (spinal) vestibular nucleus	______
11.	basal nucleus	______
12.	substantia gelatinosa	______
13.	nucleus dorsalis	______
14.	ventral tegmental nucleus	______
15.	ciliospinal nucleus	______
16.	gustatory nucleus	______
17.	magnocellular reticular nucleus	______
18.	arcuate nucleus (hypothalamus)	______
19.	central median nucleus (thalamus)	______
20.	dorsal median nucleus (thalamus)	______
21.	olfactory tubercle	______
22.	subthalamic nucleus	______
23.	lateral nucleus of accessory nerve	______
24.	nucleus intercalatus	______
25.	lower motor neurons innervating pelvic muscles	______
26.	subputaminal nucleus (substantia innominata)	______

27. a respiratory control nucleus ____________________

28. ventral nucleus of the posterior commissure ____________________

29. mesencephalic interstitial nucleus ____________________

30. dorsal tegmental nucleus ____________________

31. dorsal part of superior vestibular nucleus ____________________

32. oculomotor nuc. cells innervating medial rectus ____________________

33. paraabducens nucleus ____________________

34. ventral anterior nucleus of thalamus ____________________

35. fibers surrounding the dentate nucleus ____________________

36. spinal border cells ____________________

37. substantia nigra ____________________

38. subhypoglossal nucleus ____________________

39. lamina I of spinal dorsal horn ____________________

Neuroanatomy - Peripheral Nerves Cranial and Spinal

Directions: Indicate the name of the individual(s) in column "B" that are associated with the structure/description provided in column "A".

	A	B
1.	nervus intermedius	______________
2.	facial nerve	______________
3.	accessory nerve (superior respiratory nerve)	______________
4.	communication between ulnar and median in palm	______________
5.	accessory nerve (CN XI)	______________
6.	nasopalatine nerve	______________
7.	nerve of the pterygoid canal	______________
8.	tympanic branch of CN IX	______________
9.	auricular branch of CN X	______________
10.	middle laryngeal nerve	______________
11.	carotid sinus branch of CN IX	______________
12.	tympanic plexus	______________
13.	sinuvertebral nerve	______________
14.	submucosal plexus	______________
15.	myenteric plexus	______________
16.	first branch of lateral plantar nerve	______________
17.	endoneurium	______________
18.	intrathoracic nerve (T1 to T2 connection)	______________
19.	uterovaginal nerve plexus	______________
20.	upper trunk of brachial plexus at SCM border	______________
21.	intersegmental nodes of myelin sheath	______________
22.	musculocutaneous nerve	______________
23.	lingual branch of facial nerve	______________
24.	terminal nerve	______________

25. connection between facial and glossopharyngeal ____________________

26. supraclavicular nerves ____________________

27. aortic nerve ____________________

28. plexus of dorsal rami of C1-C3 ____________________

29. inferior cervical cardiac nerve ____________________

30. intestinal mucosal plexus ____________________

31. intracavernous plexus ____________________

32. communicating fibers between hypoglossal nerves ____________________

33. law proscribing separate fibers in dorsal and ventral roots ____________________

Neuroanatomy - Tracts

Directions: Indicate the name of the individual(s) in column "B" that are associated with the structure/description provided in column "A".

	A	B
1.	fasciculus gracilis	______________
2.	fasciculus cuneatus	______________
3.	spino-olivary tract	______________
4.	mammillo-tegmental tract	______________
5.	anterior corticospinal tract	______________
6.	dorsolateral fasciculus	______________
7.	dorsal tegmental decussation	______________
8.	ventral tegmental decussation	______________
9.	interstitio-spinal fasciculus	______________
10.	diagonal band	______________
11.	stria medullaris	______________
12.	dorsal spinocerebellar tract	______________
13.	ventral spinocerebellar tract	______________
14.	hook bundle (uncinate fasciculus of cerebellum)	______________
15.	commissure of the lateral lemniscus	______________
16.	dorsal supraoptic commissure	______________
17.	ventral supraoptic commissure	______________
18.	anterior supraoptic hypothalamic commissure	______________
19.	sulcomarginal fasciculus	______________
20.	medial lemniscus	______________
21.	septomarginal fasciculus	______________
22.	fasciculus interfascicularis (coma fasciculus)	______________
23.	solitary (respiratory)tract	______________
24.	tectospinal tract	______________
25.	rubrospinal tract	______________
26.	corticopontine tract	______________

27. fasciculus retroflexus ____________________
28. optic radiations ____________________
29. mammillo-thalamic tract ____________________
30. lenticular fasciculus ____________________
31. medial and lateral longitudinal striae ____________________
32. dorsal trigeminothalamic tract ____________________
33. superior longitudinal fasciculus ____________________
34. fasciculus solitarius ____________________
35. habenulo-interpeduncular tract ____________________
36. fronto-pontine tracts ____________________
37. lateral spinothalamic tract ____________________
38. anterior spinothalamic tract ____________________
39. septo-marginal fasciculus ____________________
40. tempro-pontine tracts ____________________
41. supramammillary decussation (commissure) ____________________
42. stria terminalis ____________________
43. cingulum bundle (fornix periphericus) ____________________
44. tectopontine tract ____________________
45. tectoreticular tract ____________________
46. dorsal longitudinal fasciculus ____________________
47. rubro-olivary-cerebellar-rubro triangle ____________________
48. olivospinal tract ____________________
49. anterolateral fasciculus ____________________
50. hippocampal commissure ____________________
51. psalterium (post part of fornix) ____________________
52. dentate nuc to red nuc to olivary nuc to cerebellum ____________________
53. spinal marginal tract ____________________
54. geniculocalcarine projections in temporal lobe ____________________
55. fasciculus proprius ____________________
56. medial longitudinal fasciculus ____________________
57. intracortical fibers in occipital lobe ____________________

58. mammillo-tegmental tract ____________________

59. prerubral fields ____________________

60. decussating cochlear fibers forming lateral lemniscus ____________________

61. sulcomarginal fasciculus ____________________

62. triangular fasciculus ____________________

63. olivocochlear bundle ____________________

64. decussation of superior cerebellar peduncle ____________________

65. limbic system circuit involved in memory ____________________

66. inter-lenticular commissure in floor of III ventricle ____________________

67. myelinated fibers terminating in medial cervical ventral horn involved in head movements ____________________

Neuroanatomy - Synonyms and Alternative Terms

Directions: Indicate the name of the individual(s) in column "B" that are associated with the structure/description provided in column "A".

	A	B
1.	ventral posterior nucleus (thalamus)	____________
2.	lateral cuneate nucleus	____________
3.	hippocampal commissure	____________
4.	pars triangularis	____________
5.	petrotympanic fissure	____________
6.	gracile tubercle	____________
7.	nucleus gracilis	____________
8.	nucleus cuneatus	____________
9.	mammillary bodies	____________
10.	superior orbital fissure	____________
11.	ciliary ganglion	____________
12.	pterygopalatine ganglion	____________
13.	facial canal	____________
14.	hypoglossal canal	____________
15.	superior colliculus	____________
16.	inferior colliculus	____________
17.	trochlear nerve	____________
18.	pneumogastric nerve	____________
19.	supracallosal gyrus	____________
20.	insular gyri	____________
21.	mammillary bodies	____________
22.	lamina terminalis	____________
23.	cavum septum pellucidum	____________
24.	trigeminal nerve	____________

25. arachnoid granulations
26. tentorial notch
27. intracerebral periarterial subarachnoid space
28. tectospinal tract
29. facial colliculus
30. solitary tract
31. fasciculus retroflexus
32. Meyer's loop
33. ligament suspends incus from roof of middle ear
34. stria medullaris in 4th ventricle
35. radial fibers of ciliary muscle
36. helicotrema
37. diploic veins
38. median raphe of the corpus callosum
39. perilymph
40. glia forming glia-pial membrane
41. endoneurium
42. cerebral aqueduct
43. vagal trigone
44. tuberculum cinereum
45. pterygoid canal
46. subcallosal gyrus
47. striopallidal axons
48. auditory (pharyngotympanic) tube
49. primary auditory cortex
50. motor language area
51. sensory language area
52. primary visual cortex
54. limbic lobe
55. anterior perforated space
56. midline invaginated part of telencephalon

57. half-moon shaped fissure capping posterolateral extent of calcarine fissure ____________________

58. amygdala ____________________

59. fusiform gyrus ____________________

60. substantia innominate ____________________

61. superior longitudinal fasciculus ____________________

62. superior fronto-occipital fasciculus ____________________

63. sleep arteries of the brain ____________________

64. inner layer of the external capsule ____________________

Microscopic Anatomy - Skin

Directions: Indicate the name of the individual(s) in column "B" that are associated with the structure/description provided in column "A".

A	B
1. tactile nerve endings found in the cornium	____________
2. tactile nerve endings in the deep layer of the epidermis	____________
3. the areolar glands of the breast	____________
4. low threshold very rapidly adapting sensory receptors	____________
5. tactile sensory nerve endings of the corium	____________
6. topical lines representing orientation of collagen fibers	____________
7. putative thermoreceptors	____________
8. spherical receptor with less thick capsule and less coiled nerve process than Krause end bulb	____________
9. cells in basal layer of epidermis containing catecholamines	____________

Microscopic Anatomy - Musculoskeletal

Directions: Indicate the name of the individual(s) in column "B" that are associated with the structure/description provided in column "A".

	A	B
1.	resorptive recesses in bone occupied by osteoclasts	________________
2.	inner or osteogenic layer of the periosteum	________________
3.	connective tissue fibers between periosteum and bone	________________
4.	canals within bone	________________
5.	opening in osteum of bone	________________
6.	storage granules of endothelial cells lining blood vessels	________________
7.	muscle fiber elevation at neuromuscular junction	________________
8.	neuromuscular junction	________________

Microscopic Anatomy - Respiratory

Directions: Indicate the name of the individual(s) in column "B" that are associated with the structure/description provided in column "A".

	A	B
1.	interalveolar pores	________________

Microscopic Anatomy - Cardiovascular

Directions: Indicate the name of the individual(s) in column "B" that are associated with the structure/description provided in column "A".

A	B
1. specialized conducting muscle cells of the heart	________________

Microscopic Anatomy - Gastrointestinal

Directions: Indicate the name of the individual(s) in column "B" that are associated with the structure/description provided in column "A".

A	B
1. splenic cords or red pulp cords	________________
2. digestive glands of the duodenum	________________
3. the phagocytic cells of the liver sinusoids	________________
4. straight, tubular intestinal glands	________________
5. eosinophilic granule-containing cells of the small intestine	________________
6. aggregations of lymphoid tissue in the small intestine	________________
7. inner layer of cells lining acinis of pancreas	________________
8. mesenchymal cells within muscle layers of the gastrointestinal tract	________________
9. splenic lymph corpuscles	________________
10. perisinusoidal spaces in liver	________________
11. liver acinus	________________

Microscopic Anatomy - Urinary

Directions: Indicate the name of the individual(s) in column "B" that are associated with the structure/description provided in column "A".

	A	B
1.	renal collecting ducts	____________
2.	renal capsule	____________
3.	urethral glands	____________
4.	paraurethral glands	____________
5.	peri-renal fascia	____________
6.	interureteric ridge	____________
7.	retro-renal fascia	____________
8.	renal columns	____________
9.	glomerular capsule	____________
10.	sphincter urethrae	____________
11.	urinary trigone	____________
12.	urethral lacunae	____________
13.	"U" shaped part of the nephron	____________
14.	watershead between anterior and posterior branches of the renal arteries	____________
15.	convoluted tubule of the kidney	____________
16.	rete testes	____________
17.	renal pyramid	____________

Microscopic Anatomy - Reproductive

Directions: Indicate the name of the individual(s) in column "B" that are associated with the structure/description provided in column "A".

A	B
1. testosterone-producing cells of the testes	________
2. cells supporting the testicular epithelium	________
3. straight tubules of testes	________
4. coiled part of efferent ductules of testes	________

Microscopic Anatomy - Nervous System

Directions: Indicate the name of the individual(s) in column "B" that are associated with the structure/description provided in column "A".

	A	B
1.	largest cell of the cerebellar cortex	______
2.	myelin producing cell in the PNS	______
3.	rough endoplasmic reticulum	______
4.	flattened membranous cisternae in nerve cell body	______
5.	small bundles of unmyelinated fibers in a peripheral nerve	______
6.	large pyramidal cells of the motor cortex	______
7.	junctions between adjacent Schwann cells	______
8.	incisures in peripheral myelinated nerve fibers	______
9.	radial glia	______
10.	feathered glial cells	______
11.	endoneurium	______
12.	large pyramidal cells in calcarine cortex	______
13.	longitudinal Schwann cells in re-growing axons	______
14.	inhibitory interneurons in spinal cord	______
15.	ADH and oxytocin bodies in posterior pituitary	______
16.	small horizontal cells in the cerebral cortex	______
17.	small rounded multipolar cells on cerebral cortex	______
18.	visual cortex laminae III	______
19.	visual cortex "solitary" cells	______
20.	visual cortex "star pyramids)	______

Microscopic Anatomy - Eye and Ear

Directions: Indicate the name of the individual(s) in column "B" that are associated with the structure/description provided in column "A".

	A	B
1.	apocrine glands of the eyelid	________________
2.	sebaceous glands of the eye lashes	________________
3.	tarsal glands of the eyelid	________________
4.	scleral venous sinus	________________
5.	membrane between corneal epithelium and stroma	________________
6.	membrane between corneal endothelium and stroma	________________
7.	vestibular membrane of the cochlea	________________
8.	structure containing auditory receptor cells	________________
9.	spiral canal between inner and outer pillar cells	________________
10.	columnar cells on floor of cochlear duct	________________
11.	optic disc (blind spot)	________________

Microscopic Anatomy - Endocrine

Directions: Indicate the name of the individual(s) in column "B" that are associated with the structure/description provided in column "A".

	A	B
1.	insulin-producing, cells in the pancreas	________________
2.	inner layer of pancreatic acinus	________________

Developmental Anatomy - Embryology

Directions: Indicate the name of the individual(s) in column "B" that are associated with the structure/description provided in column "A".

	A	B
1.	sinus venosus	______________
2.	ovarian follicle	______________
3.	paramesonephric duct	______________
4.	cartilage of the 2nd branchial arch	______________
5.	embryonic mesonephric duct	______________
6.	ductus venosus	______________
7.	remnants of Wolffian duct in paroophoron	______________
8.	embryonic urethral glands	______________
9.	ductus arteriosus	______________
10.	longitudinal duct of the epoophoron	______________
11.	epoophoron	______________
12.	cartilage of the 1st branchial arch	______________
13.	fissure between subarachnoid space and middle ear	______________
14.	neurenteric canal or blastoporic canal	______________
15.	foramen ovale	______________
16.	ostium secundum	______________
17.	primitive node (knot)	______________
18.	ectodermal source of anterior pituitary gland	______________
19.	ganglion cells on CN XII	______________
20.	terminal junction of the cardinal veins	______________
21.	oral ectodermal pouch forming anterior pituitary	______________
22.	diverticulum of the ileum	______________
23.	gelatinous connective tissue of the umbilical cord	______________
24.	remnant of Wolffian body at posterior side of the testicle	______________
25.	canal in the fetal sphenoid bone	______________

26. mesonephric duct ____________________

27. paraumbilical veins ____________________

28. remnant of the mesonephric duct ____________________

29. ovarian follicle ____________________

Section Two - Answer Keys

Gross Anatomy - Head and Neck

Directions: Indicate the name of the individual(s) in column "B" that are associated with the structure/description provided in column "A".

	A	B
1.	maxillary sinus	Highmore
2.	facial canal	Fallopi
3.	lesser (small) sublingual glands	Rivinus
4.	posterolateral pharyngeal recess	Rosenmuller
5.	pterygoid canal and its vessels	Vidian
6.	pharyngeal lymphatic ring	Waldeyer
7.	submandibular duct	Wharton
8.	cuneiform cartilage of larynx	Wrisberg/ Morgagni
9.	opening into cavernous sinus for passage of VI nerve	Dorello
10.	smooth muscle of eyelids	Mueller
11.	parotid duct	Stenson / Blasius
12.	pharyngo-tympanic tube	Eustatio
13.	tarsal glands of the eyelid	Meibomian
14.	fascial sheath of the eye	Tenon
15.	laryngeal ventricle	Hilton
16.	aryepiglottic muscle	Hilton
17.	superior nasal concha	Morgagni
18.	accessory lacrimal glands	Wolfring / Krause Baumgarten / Henle
19.	petroclinoid ligament	Gruber
20.	carotid tubercle	Chassaignac
21.	vomeronasal cartilage	Jacobson
22.	corniculate cartilage of the larynx	Santorini

23.	space between dural folds that houses trigeminal ganglion	Meckel
24.	temporo-squamoual suture	Glassenman
25.	suspensory ligament of the lens	Zinn
26.	cleft in meningeal dura housing trigeminal ganglion	Meckel
27.	cartilage of the 2nd branchial arch	Reichert
28.	small bones found in sutures of the skull	Worm
29.	canal in temporal bone for auricular branch of X	Arnold
30.	columns in the osseous spiral lamina of the cochlea	Cotunnius
31.	foramen cecum of tongue	Morgagni
32.	space between the upper border of levator veli palatini and base of the skull	Morgagni
33.	tympanic notch on tympanic part of temporal bone	Rivinus
34.	palpebral portion of the lacrimal gland	Rosenmuller
35.	incisive foramina of the hard palate	Stenson
36.	venous sinus of the sclera	Schlemm / Lauth
37.	lacrimal fold	Hasner / Huschke
38.	ductus reuniens	Henson
39.	thickened lower part of the bulbar sheath	Lockwood
40.	fold on posterior pharyngeal wall	Passavant
41.	tensor tarsi (lateral part of orbicularis oculi)	Horner
42.	small swelling on posterior helix of ear	Darwin / Woolner
43.	lowest thyropharyngeus fibres above the cricopharyngeus muscle	Killian
44.	five artery anastomoses on anterior nasal septum	Little / Kiesselbach
45.	line from inferior margin of orbit to external auditory canal	Reid
46.	vein connecting pterygoid plexus and cavernous sinus	Vesalius
47	membrane fold in lacrimal duct near punctum	Botchdalek / Foltz
48.	retropharyngeal lymph nodes	Rouvier

49.	tympanic canal	Jacobson
50.	sublingual gland duct	Bartholin
51.	sphenoid conchae	Bertin
52.	space above jugular notch	Plender
53.	small defect on posterior edge of tympanic ring	Rininus
54.	tubular cavity in nasal septum	Ruysch
55.	accessory lacrimal glands	Ciaccio / Wolfring Baumgarten / Henle
56.	deep lingual arterial anastomosis	Beclard
57.	space in iridocorneal angle	Fontana
58.	thyroglossal duct	Bochdalek / Hiss Vater / Sappy
59.	buccal fat pad	Bichat
60.	anterior process of malleus	Rau
61.	helicotrema	Breschat
62.	opening in osseous part of external auditory meatus	Huschke
63.	minor sublingual glands	Walther / Rivinus
64.	petrotympanic fissure	Glaser
65.	scaphoid fossa in sphenoid bone	Cruveilhier
66.	ciliary muscle	Bowman
67.	membrane lining the maxillary sinus	Schneider
68.	alar ligament	Mauchart
69.	parietal emissary veins	Santorini
70.	lacrimal part of orbicularis oculi	Duverney
71.	triangle involving hyoglossus, digastric and hyoid bone	Beclard
72.	pharyngeal ridge with contraction of superior constrictor	Passavant
73.	pyramidal lobe of thyroid gland	Zuckerkandl
74.	sublingual bursa	Fleischmann
75.	muscle fibers in check ligament of the eye	Sappey

76.	tentorial notch	Pacchioni
77.	central artery of the retina	Zinn
78.	lateral capsular ligament of temporomandibular joint	Ferrein
79.	hiatus for the lesser petrosal nerve	Arnold
80.	thyroid ima artery	Neubauer
81.	auricular ligaments	Valsalva
81.	hiatus for the greater petrosal nerve	Falopii / Ferein
		Torin
82.	anterior lingual gland	Bauhin / Blandin
		Nuhn
83.	laryngeal ventricle	Morgagni
84.	meridional fibers of the ciliary muscle	Brucke / Crampton
85.	suspensory ligament of the thyroid gland	Berry
86.	tubal tonsil	Gerlach
87.	cartilage attached to vocal process of arytenoid cartilage	Seiler
88.	risorius muscle	Santorini
89.	line passing between opisthion and basion	Daubenton
90.	sphenoidal concha	Bertin
91.	depression in lacrimal sac for lacrimal duct	Maier
92.	subarachnoid space around optic nerve	Schwalbe
93.	petro-occipital fissure	Ecker
94.	parotid duct	Blasius
95.	optic disc	Mariotte
96.	scalenus minimus	Sibson
97.	spiral canal of cochlea	Rosenthal
98.	diploic veins	Dupytren / Breschet
99.	olfactory glands	Bowman
100.	innermost layer of the choroid of the eye	Bruch
101.	ligament between pars tensa and pars flaccida	Troltsch
102.	small recess in middle ear cavity	Prussak

103.	true vocal folds	Ferrein
104.	retrolingual space	Bogros
105.	gap in petrotympanic fissure for chorda tympani	Huger / Civinini
106.	highest nasal concha	Santorini / Morgagni
107.	ligament of the malleus	Casserio
108.	superficial portion of orbicularis oculi	Riolan
109.	small mucous gland in floor of the mouth	Suzanne
110.	fovea centralis	Sommering
111.	clivus	Blumenbach
112.	pretarsal segment of orbicularis oculi	Horner
113.	endolymphatic sac	Bottcher
114.	lateral tubercle of the orbit	Whitnall
115.	mastoid air sinuses	Valsalva
116.	anterior part of fascial sheath of the eye	Bonnet
117.	suspensory ligament of the esophagus	Gillette
118.	parathyroid gland	Sandstrom
119.	mastoid fontanelle	Casserio
120.	pterygopalatine fossa	Bichat
121.	small cavity on floor of external auditory canal	Meyer
122.	small cartilage on anterior fold of vocal cord	Luschka
123.	bursa posterior to hyoid bone	Boyer
124.	superior transverse ligament of the eye	Whitnall
125.	potential space between ciliary zonule and vitreous body	Hannover
126.	anterior limiting lamina of cornea	Schwalbe
127.	notches on cartilage of external auditory meatus	Santorini
128.	conjunctival glands	Terson
129.	sphenoparietal sinus	Breschet / Scarpa
130.	pterygospinous ligament and process	Civinini
131.	carotid space (triangle)	Malgiagne
132.	cochlear duct	Lowenberg

133.	pharyngeal tonsil	Luschka
134.	small orbital muscle that pulls eye forward	Landstrom
135.	short ciliary arteries on sclera near optic nerve	Haller
136.	anterior tympanic artery	Glaser
137.	taste bud	Schwalbe
138.	pars flaccida	Shrapnell
139.	horizontal cartilage near nasal septum	Huschke
140.	cutaneous muscle of the lip	Aeby / Klein
142.	endolymphatic recess	Hyrtl
143.	space containing carotid bifurcation	Gerdy
144.	tonsillar fossa	Tortual
145.	episcleral space	Tenon
146.	connective tissue fibers bounding pars flaccida	Prussak
147.	triangle involving digastric, mylohyoid and hypoglossal nerve	Piragov
148.	median atlanto-occipiral ligament	Cruveilhier
149.	tympano-meningeal fissure	Hyrtl
150.	ansa subclavian	Vieussens
151.	epithelial tissue near angle of mandible by the buccal nerve	Chievitz
152.	bulb of the jugular vein	Heister
153.	canal in vitreus from optic disc to lens	Cloquet / Stilling
154.	small venous plexus between canal of Schlemm and space of Fontana	Leber
155.	space between tongue and sublingual gland	Sebileau
156.	exocrine salivary glands	von Ebner
157.	arterial plexus of the nasal septum	Kiesselbach / Little
158.	suprameatal triangle / mastoid fossa	McEwan
159.	left supraclavicular lymph node	Virchow
160.	triangular area on posterior proximal esophagus	Laimer

Gross Anatomy - Limbs

Directions: Indicate the name of the individual(s) in column "B" that are associated with the structure/description provided in column "A".

	A	B
1.	iliofemoral ("Y") ligament	Bigelow / Bertin
2.	adductor canal	Hunter
3.	dorsal tubercle of the radius	Lister
4.	inguinal ligament	Poupart
5.	posterior meniscofemoral ligament	Wrisberg
6.	supinator arch	Frohse
7.	fibula-talo-calcaneal ligament	Rouviere
8.	iliopectineal fascia	Cooper
9.	line joining ASIS to ischial tuberosity	Nelaton
10.	common tendon of gastrocnemius and soleus	Achilles
11.	oblique popliteal ligament	Bourgery
12.	transverse tarsal joint	Chopart
13.	palmar fascia of the hand	Dupytrens
14.	superior glenohumeral ligament	Flood
15.	suspensory ligament of the axilla	Gerdy
16.	anterior meniscofemoral ligament	Humphrey
17.	tarso-metatarsal joint	Lisfranc
18.	iliotibial tract (band)	Maissiat
19.	femoral triangle	Scarpa
20.	oblique cord of proximal radio-ulnar joint	Weitbrecht
21.	abnormal abductor digiti minimi of the hand	Wood
22.	tunnel transmitting the ulnar nerve at the wrist	Guyon
23.	ligament forming tunnel through which passes the median nerve and brachial artery	Struthers
24.	cutaneous ligaments of the digits	Cleland

25.	transverse band of the ulnar collateral ligament	Cooper
26.	canals within bones containing vessels	Havers
27.	lacunae within bones containing osteoclasts	Howship
28.	inguinal lymphatic nodes adjacent to femoral canal	Rosenmuller
29.	lateral tubercle of tibia for attachment of the iliotibial band	Gerdy
30.	angle of inclination of the femur	Alberg
31.	posterior process of the talus (astragulus)	Stieda
32.	lymph node at the femoral ring	Cloquet
33.	inferior calcaneal nerve	Baxter
34.	posterior process of the talus	Stieda
35.	fibrous roof over cubital tunnel	Osborne
36.	lower fibers of posterior sacroiliac joint	Bichat
37.	articularis genu muscle	Dupre
38.	proximal of the two palmar limb creases	Sydney
39.	coracobrachialis muscle	Casserio
40.	semimembranosus bursa	Brodie
41.	ligament from posterior superior iliac spine to 2nd transverse tubercle of sacrum	Zaglas
42.	calcar femorale	Bigelow
43.	lacertus fibrosus	Pirogov
44.	articular cartilage of the acromioclavicular joint	Weitbrecht
45.	dorsal talonavicular ligament	Chopart
46.	tuberosity of the 5th metatarsal	Vesalius
47.	fibular collateral ligament	Winslow
48.	septal tissue in femoral canal	Cloquet
49.	supraclavicular lymph nodes on left side	Virchow
50.	superficial transverse ligament of the palm	Gerdy
51.	accessory anterior inferior talofibular ligament	Bassett
52.	plantar ligament of the foot	Cruveilhier
53.	line passing through great toe and mid calcaneous	Meyer

54.	split in flexor digitorum superficialis tendon for tendon of flexor digitorum profundus	Camper
55.	longitudinal fibers of articular capsule of hip joint	Weitbrecht
56.	volar carpal ligament	Mayer
57.	space between pronator quadratus and flexor digitorum profundus in the forearm	Parona
58.	location of posterior tibial artery behind medial malleolus	Pimenta

Gross Anatomy - Thorax

Directions: Indicate the name of the individual(s) in column "B" that are associated with the structure/description provided in column "A".

	A	B
1.	suspensory ligament of the breast	Cooper
2.	sternal angle	Louis
3.	sebaceous glands of the areola	Montgomery
4.	conduction fibers of the heart	Purkinje
5.	suprapleural membrane (fascia)	Sibson
6.	valve of the coronary sinus	Thebesia
7.	atrio-ventricular bundle	His / Kent / Paladino
8.	attachment for anterior scalene on the first rib	Lisfranc
9.	long thoracic nerve	Bell
10.	foramen ovale of the heart	Botallo
11.	ductus (ligamentum) arteriosum	Botallo / Harvey
12.	space above the jugular notch	Burns
13.	sino-atrial node	Flack / Keith / Koch
14.	oblique vein of the left atrium	Marshall
15.	smooth muscle fibers of the smallest bronchi	Reisseisen
16.	atrio-ventricular node	Aschoff-Tawara
17.	retromammary space	Chassaignac
18.	valve of the inferior vena cava	Eustatio
19.	valve of the coronary sinus	Thebesia
20.	fold in left superior vena cava in pericardial sac	Marshall
21.	third intercostal space on the left sternal border	Erb
22.	lymphatics in areola of nipple	Sappey
23.	sternopericardial ligament	Luschka / Lanneulongue
24.	partial duplication of thoracic duct in thorax	Haller
25.	nodules on edge of semilunar valves	Arantis / Morgagni
26.	interspinous plane	Lanz

27.	thoracic duct	van Hoorne / Pecquet
28.	transverse pericardial sinus	Theile
29.	triangular area in wall of right atrium	Koch
30.	pulmonary ridge	Mall
31.	lacteal veins	Aselli
32.	innominate cardiac vein	Vieussen
33.	mammary tissue extending into axilla	Spence
34.	anastomosis between circumflex a. and R coronary	Kugel
35.	largest openings of the small cardiac veins	Lannelongue / Vieussen Thebesius
36.	pulmonary circulation	Servetus
37.	space between costal and sternal attachments of diaphragm	Larrey / Morgagni
38.	intervenous tubercle of right ventricle	Lower
39.	infraclavicular fossa	Mohrenheim
40.	small air filled sacs on surface of lung	Malpighi
41.	arterial conduction system of the heart	Bachmann
42.	interpectoral lymph nodes	Rotter

Gross Anatomy - Abdomen

Directions: Indicate the name of the individual(s) in column "B" that are associated with the structure/description provided in column "A".

	A	B
1.	main pancreatic duct	Wirsung
2.	accessory pancreatic duct	Santorini
3.	hepatorenal recess	Morrison
4.	epiploic foramen	Winslow / Duverney
5.	lacunar ligament	Gimbernat
6.	anterior renal capsule (fascia)	Gerota
7.	spiral fold of cystic duct	Heister
8.	inguinal triangle	Hesselbach
9.	hepatopancreatic sphincter	Oddi / Glisson
10.	inferior lumbar triangle	Petit
11.	lymphatic nodules of the ileum	Peyer
12.	suspensory ligament of the duodenum	Treitz
13.	hepatopancreatic ampulla	Vater
14.	fatty layer of superficial abdominal fascia	Camper
15.	fibrous layer of superficial abdominal fascia	Scarpa
16.	ileal diverticulum	Meckel
17.	insulin secreting cells of pancreas	Langerhans
18.	iliocecal fold	Treves
19.	capsule of the liver	Glisson
20.	caudate lobe of the liver	Spiegel
21.	cisternal chyli	Pecquet
22.	supraduodenal artery	Wilkie
23.	mucosal membrane folds in the ileum	Leiberkuhn
24.	vein crossing pyloduodenal junction	Mayo
25.	sternocostal triangle (foramen)	Morgagni
26.	lumbocostal trigone (foramen)	Bochdalek

27.	posterior renal capsule (fascia)	Zuckerkandl
28.	transpyloric plane	Addison
29.	pudendal canal	Alcock
30.	iliofemoral ligament	Bigelow
31.	capsule surrounding the glomerulus of the kidney	Bowman
32.	glands secreting digestive enzymes in duodenum	Brunner / Wepfer
33.	upper part of the pectineal fascia	Cooper
34.	dilatation at the neck of the gall-bladder	Hartman
35.	"u" shaped region of the nephron	Henle
36.	circular mucous folds of the small intestine	Kerkering
37.	inguinal ligament	Poupart
38.	ileocecal fold	Treves
39.	ileocecal valve	Tulp / Brauhin / Varolio
40.	reflected part of the inguinal ligament	Colles
41.	linea semilunaris	Douglas
42.	spiral valves of the cystic duct	Amussat
43.	tubular structures in the wall of the gallbladder	Douglas
44.	conjoint tendon of the transversus abdominis muscle	Henle
45.	small tubular structures in the wall of the gallbladder	Luschka
46.	renal collecting ducts	Bellini
47.	perirenal fascia	Colles
48.	retroaortic left renal vein	Schnitker
49.	cysto-hepatic triangle	Calot
50.	small veins of the falciform ligament	Sappey
51.	semilunar line of the abdominal wall	Spiegel
52.	line connecting gall bladder and IVC on ventral liver	Cantile
53.	uncovertebral joints	Luschka
54.	surface location for inflamed appendix	McBurney
55.	ischioanal fossa	Chaussier
56.	ischiorectal fossa	Velpeau

57.	area bounded by lower border L lung, ant border of spleen, L costal margin, inf margin L lobe of liver	Traube
58.	superior hypogastric plexus	Latarget
59.	small cutaneous veins around umbilicus	Sappey
60.	ileal sphincter	Varolio
61.	central anastomotic mesenteric artery	Riolan / Moskowitz
62.	middle lobe of prostate gland	Morgagni
63.	mucous membrane at base of appendix	Gerlach
64.	celiac trunk	Haller
65.	arterial anastomosis between SMA and IMA	Drummond
66.	fissure on liver between right lobe and caudate lobe	Rouviere
67.	esophagogastric angle	His
68.	paraaortic bodies	Zuckerkandl
69.	deep inguinal lymph nodes	Cloquet / Eglis Rosenmuller
70.	central veins of the liver	Krukemberg
71.	small mucous glands of the biliary mucosa	Theile
72.	uncinated process of the pancreas	Winslow
73.	frenulum of the ileal orifice	Morgagni
74.	intestinal glands	Galeati / Lieberkuhn
75.	capsule of the spleen	Malpighi
76.	inguinal canal	Velpeau / Vesalius
77.	interfoveolar ligament	Hesselbach
78.	tongue of right lobe of liver extending downward	Riedel
79.	circular membranous folds of small intestine	Kerkering
80.	superior lumbar triangle	Grynfeltt-Lesshaft
81.	precolic fascia	Jackson
82.	iliopubic ligament	Thompson (H)
83.	retroperitoneal veins	Retzius / Ruysch
84.	falciform margin of saphenous opening	Hey
85.	line between ASIS and umbilicus	Monro-Richter

86.	mesenterico-parietal fascia	Waldeyer
87.	transversalis fascia	Hyrtl
88.	anterior layer of rectus sheath below arcuate line	Spieghel
89.	inferior duodenal fossa/recess	Gruber-Landzert
90.	superior mesenteric artery	Wilkie
91.	esophagogastric angle	His
92.	intrahepatic bile ductules	Hering
93.	sling-like collar at gastroesophageal junction	Helvetius

Gross Anatomy - Pelvis and Perineum

Directions: Indicate the name of the individual(s) in column "B" that are associated with the structure/description provided in column "A".

	A	B
1.	anal columns	Morgagi
2.	perineal fascia	Colles / Cruveilhier
3.	bulbourethral glands	Cowper / Mery
4.	pudendal canal	Alcock
5.	deep fascia of the penis	Buck
6.	rectouterine (rectovesical) pouch	Douglas
7.	transverse rectal folds	Houston
8.	urethral glands	Littre / Guerin
9.	persistant processus vaginalis in females	Nuck
10.	retro-pubic space	Retzius
11.	paraurethral glands	Skene
12.	greater vestibular glands	Bartholin /Tiedmann Duverney
13.	fascia separating prostate from rectum	Devonvilliers
14.	external sphincter of the urethra	Guthrie / Wilson J
15.	cardinal ligament of the uterus	Mackenrodt
16.	inter-ureteric ridge or crest	Mercier
17.	trigone of the urinary bladder	Lieutaud
18.	rectosacral fascia	Waldeyer
19.	rectal valves	Ball
20.	uterine tubes	Falloppio
21.	perineal body	Savage
22.	anococcygeal body	Symington
23.	iliopectineal fascia	Thompson
24.	ampulla of the ductus deferens	Henle
25.	space in the bulbous portion of the urethra	LeCat

26.	enlarged middle lobe of the prostate gland	Morgagni / Home
27.	openings in the mucous glands of the male urethrae	Morgagni
28.	circular muscle in the fundus uterus	Ruysch
29.	retro-renal fascia	Zuckerkandl
30.	watershead between anterior and posterior branches of the renal artery	Bdodel
31.	deep perineal fascia	Gallaudet
32.	lateral edge of trigone area of urinary bladder	Bell
33.	coccygeal body	Luschka
34.	nerve ganglion in lateral wall of cervix	Frankenhauser
35.	paradidymis	Giraldes
36.	remnant of peritoneal processus vaginalis	Haller
37.	cremaster muscle	Riolan
38.	mucous secreting glands lateral to lower vagina	Duverney
39.	superficial fascia of the perineum	Cruveilhier
40.	lobules of epididymis	Haller
41.	subperitoneal fascia anterior to external iliac artery	Abernahy
42.	transverse line marking sacral promontory	Kilian
43.	perineal membrane	Camper
44.	puboprostatic ligament	Denonvillier
45.	navicular fossa of urethra	Morgagni
45.	longitudinal muscle of rectal wall	Kohlrausch
46.	deep fibers of dartos muscle in scrotal septum	Sebileau
47.	ovarian fossa	Claudius
48.	perianal lymphatic plexus	Quenu
49.	prostatic venous plexus	Santorini
50.	cremasteric fascia	Cooper / Scarpa
51.	rectovaginal septum	Tyrrell
52.	round ligament of the uterus	Hunter
53.	ischioanal fossa	Velpeau
54.	connective tissue fat in labia majora	Broca

55.	mesentery from broad ligament to appendix	Clado
56.	fundiform ligament of the clitoris	Retzius
57.	paraurethral ducts	Skene / Schuller
58.	scrotal raphe	Vesling
59.	valve of the navicular fossa	Guerin
60.	middle transverse rectal fold	Kohlrausch
61.	uvula of the bladder	Lieutaud
62.	posterior urethral valve	Amussat
63.	suspensory ligament of the ovary	Clado
64.	space between urinary bladder and intramural part of ureter	Waldeyer
65.	smegma secreting glands of penis	Tyson
66.	superficial perineal space	Colles
67.	gubernaculum testes	Hunter
68.	lower 1/3 of the anal canal	Straud
69.	frenulum of ileocecal valve	Morgagni
70.	transverse perineal ligament	Krause, K
71.	isthmus of the ureter	Guyon, F
72.	connective tissue between bladder and anterior vaginal wall	Halban
73.	folds of peritoneum containing rectouterine muscle	Petit, A

Neuroanatomy - Brain and Spinal Cord

Directions: Indicate the name of the individual(s) in column "B" that are associated with the structure/description provided in column "A".

	A	B
1.	cerebral aqueduct	Sylvius
2.	lateral aperture of the IV ventricle	Luschka
3.	midline aperture of the IV ventricle	Magendie
4.	reticular process of the cord (lateral horn)	Lenhossek
5.	ventriculus terminalis (of the spinal cord)	Krause
6.	foramen cecum (pons/medullary junction)	Vic d'Azyr
7.	tuberculum cinereum	Rolando
8.	arachnoid from dorsum sellae to mammillary body	Liliequist
9.	interpeduncular fossa	Tarin
10.	small depression just anterior to obex	Arantii
11.	oligodendrocyte/Schwann cell border	Obersteiner-Redlich
12.	hippocampus	Ammon
13.	interventricular foramen	Monro
14.	cerebellar granular layer	Turner
15.	superior medullary velum	Vieussens
16.	cavum septum pellucidum	Duncan
17.	pons	Varoli
18.	microglia	Hortega
19.	lamina I of spinal cord	Waldeyer
20.	substantia gelatinous	Rolando
21.	tapetum	Fielding and Rell
22.	anterior part of the III ventricle	Wilder
23.	superior anastomotic vein	Trolard
24.	inferior anastomotic vein	Labbe'
25.	great cerebral vein	Galen
26.	confluence of sinuses	Herophilus

27.	arterial circle at base of brain	Willis
28.	large spinal artery at T9 –T12 level	Adamkiewicz
29.	cortical areas designated by number	Brodmann
30.	expressive language area	Broca
31.	receptive language area	Wernicke
32.	substantia innominata	Meynert
33.	laminae of white fibers in cerebral cortex	Baillarger
34.	stria of the fourth ventricle	Bergmann
35.	quadrigeminal cistern	Bichat
36.	paraolfactory area	Broca
37.	olfactory islands	Calleja
38.	tail of the dentate gyrus	Giacomini
39.	pontocerebellar cistern	Hilton
40.	reticular formation	Lenhossek
41.	olfactory bulb	Morgagni
42.	insular lobe	Reil
43.	trigone of the lateral lemniscus	Reil
44.	tuberculum cinerium	Rolando
45.	cerebral basal vein	Rosenthal
46.	cistern of the lateral fissure	Sylvius
47.	inferior medullary velum	Tarin
48.	centrum semiovale	Vic d'Azyr / Vieussen
49.	artery of the pterygoid canal	Vidian
50.	cavity of the septum pellucidum	Vieussen / Sylvius
51.	subarachnoid space around cerebral vessels	Virchow – Robbins
52.	intercavernous (circular) sinus	Ridley
53.	arachnoid granulations	Pacchioni
54.	terminal ventricle of the spinal cord	Krause
55.	calamus scriptorus	Arantius
56.	linea splendens	Haller
57.	single artery from PCA supplying thalamus and	

	midbrain bilaterally	Percheron
58.	band of the dentate gyrus	Giacomini
59.	substantia reticularis alba	Arnold
60.	cornu ammonis	Ammon
61.	arbor vitae (of cerebellum)	Willis

Neuroanatomy - Sulci and Gyri

Directions: Indicate the name of the individual(s) in column "B" that are associated with the structure/description provided in column "A".

	A	B
1.	central sulcus	Rolando
2.	horizontal fissure	Sylvius
3.	superior transverse temporal gyrus	Heschl
4.	left middle frontal gyrus	Exner
5.	rhinal sulcus	Retzius
6.	hypothalamic sulcus	Monro
7.	gyrus descendens (lateral occipital lobe)	Ecker
8.	circular sulcus	Reil
9.	supraorbital sulcus	Broca
10.	incisura temporalis	Schwalbe
11.	gyrus olfactorio-orbitalis	Retzius
12.	subcallosal gyrus	Zuckerkandl
13.	dentate gyrus	Huxley
14.	supracallosal gyrus (induseum griseum)	Retzius
15.	sulcus intermedius secundus	Eberstaller
16.	intraparietal sulcus	Turner
17.	insular cortex	Reil
18.	gyrus lateral to fascia dentate	Anders-Retzius
19.	fissure below splenium of corpus callosum	Bichat
20.	occipital angle of the parietal lobe	Broca
21.	medial olfactory gyrus	Retzius
22.	paraolfactory area	Broca
23.	hippocampus	Ammon
24.	transverse gyrus of the insula	Eberstaller
25.	descending occipital sulcus	Echer
26.	inferior frontal gyrus	Broca

27.	sulcus from lower central sulcus to occipital lobe	Pansch
28.	subsplenial gyrus	Anders-Retzius
29.	diagonal band	Broca
30.	gyrus fornicatus	Meynert
31.	transverse parietal sulcus	Brissaud
32.	first intermediate sulcus (parietal lobe)	Jensen
33.	second intermediate sulcus	Eberstaller
34.	anterior transverse temporal sulcus	Schwalbe
35.	cingulate sulcus	Broca
36.	parietooccipital sulcus	Gratiolet
37.	preoccipital notch	Meynert
38.	diagonal sulcus (frontal lobe)	Eberstaller
39.	radiate sulcus (frontal lobe)	Eberstaller
40.	frontomarginal sulcus	Wernicke
41.	transverse insular sulcus	Eberstaller
42.	parietooccipital arc	Gratiolet
43.	pars triangularis (frontal lobe)	Broca
44.	pars opercularis	Broca
45.	superior temporal gyrus (posterior part)	Wernicke

Neuroanatomy - Ganglia

Directions: Indicate the name of the individual(s) in column "B" that are associated with the structure/description provided in column "A".

	A	B
1.	trigeminal ganglion	Gasser / Caccerio
2.	pterygopalatine ganglion	Meckel
3.	petrosal (inferior) ganglion of IX	Andersch
4.	vestibular ganglion	Scarpa
5.	jugular (superior) ganglion of IX	Ehrenritter
6.	cells of the sympathetic ganglia	Dogiel
7.	otic ganglion	Arnold
8.	spiral ganglion	Corti
9.	rudimentary ganglion cells on CN XII	Froriep
10.	submandibular ganglion	Langley
11.	sympathetic cells near coronary sinus entry into atrium right atrium	Remak
12.	cardiac ganglia	Wrisberg
13.	celiac ganglion	Vieussen
14.	myenteric ganglia of gut	Aurbach
15.	submucosal ganglia of gut	Meissner
16.	cochlear ganglion	Bottcher
17.	carotid ganglion	Laumonier
18.	autonomic ganglia of the stomach	Remak
19.	ganglion impar	Walther
20.	ganglion of the superior alveolar nerve	Valentin

Neuroanatomy - Nuclei

Directions: Indicate the name of the individual(s) in column "B" that are associated with the structure/description provided in column "A".

	A	B
1.	nucleus gracilis	Gall
2.	nucleus cuneatus	Burdach
3.	midbrain parasympathetic nucleus	Edinger-Westphal
4	ventral posterior medial nucleus of thalamus	Flechsig
5.	lateral (external) cuneate nucleus	Monakow
6.	interpeduncular nucleus	Gudden
7.	superior vestibular nucleus	Bechterew
8.	medial vestibular nucleus	Schwalbe
9.	lateral vestibular nucleus	Deiter
10.	inferior (spinal) vestibular nucleus	Roller
11.	basal nucleus	Meynert
12.	substantia gelatinosa	Rolando
13.	nucleus dorsalis	Le Gros Clark
14.	ventral tegmental nucleus	Tsai
15.	ciliospinal nucleus	Budge
16.	gustatory nucleus	Nageotte
17.	magnocellular reticular nucleus	Kolliker
18.	arcuate nucleus (hypothalamus)	Kolliker
19.	central median nucleus (thalamus)	Luys
20.	dorsal median nucleus (thalamus)	Burdach
21.	olfactory tubercle	Ganser
22.	subthalamic nucleus	Luys
23.	lateral nucleus of accessory nerve	Roller
24.	nucleus intercalatus	Staderini
25.	lower motor neurons innervating pelvic muscles	Onuf
26.	subputaminal nucleus (substantia innominata)	Meynert / Reichert / Ayola

27.	a respiratory control nucleus	Kolliker-Fuse
28.	ventral nucleus of the posterior commissure	Darkschewitsch
29.	mesencephalic interstitial nucleus	Cajal
30.	dorsal tegmental nucleus	Gudden
31.	dorsal part of superior vestibular nucleus	Onufrowicz
32.	oculomotor nuc. cells innervating medial rectus	Perlia or Spitzka
33.	paraabducens nucleus	Fuse
34.	ventral anterior nucleus of thalamus	Hassler
35.	fibers surrounding the dentate nucleus	Stilling
36.	spinal border cells	Stilling
37.	substantia nigra	Soemmering
38.	subhypoglossal nucleus	Staderini
39.	lamina I of spinal dorsal horn	Waldeyer

Neuroanatomy - Peripheral Nerves Cranial and Spinal

Directions: Indicate the name of the individual(s) in column "B" that are associated with the structure/description provided in column "A".

	A	B
1.	nervus intermedius	Wrisberg
2.	facial nerve	Bell
3.	accessory nerve (superior respiratory nerve)	Bell
4.	communication between ulnar and median in palm	Benettini
5.	accessory nerve (CN XI)	Willis
6.	nasopalatine nerve	Cotunnius
7.	nerve of the pterygoid canal	Vidian
8.	tympanic branch of CN IX	Jacobson
9.	auricular branch of CN X	Arnold
10.	middle laryngeal nerve	Exner
11.	carotid sinus branch of CN IX	Hering
12.	tympanic plexus	Jacobson
13.	sinuvertebral nerve	Luschka
14.	submucosal plexus	Meissner
15.	myenteric plexus	Auerbach
16.	first branch of lateral plantar nerve	Baxter
17.	endoneurium	Henle or Key and Retzius
18.	intrathoracic nerve (T1 to T2 connection)	Kuntz
19.	uterovaginal nerve plexus	Frankenhauser
20.	upper trunk of brachial plexus at SCM border	Erb
21.	intersegmental nodes of myelin sheath	Ranvier
22.	musculocutaneous nerve	Casserio
23.	lingual branch of facial nerve	Cruveilhier
24.	terminal nerve	Fritsch

25.	connection between facial and glossopharyngeal	Haller
26.	supraclavicular nerves	Langenbeck
27.	aortic nerve	Ludwig
28.	plexus of dorsal rami of C1-C3	Cruveilhier
29.	inferior cervical cardiac nerve	Pavlov
30.	intestinal mucosal plexus	Remak
31.	intracavernous plexus	Walther
32.	communicating fibers between hypoglossal nerves	Hyrtl
33.	law proscribing separate fibers in dorsal and ventral roots	Bell-Magendie

Neuroanatomy - Tracts

Directions: Indicate the name of the individual(s) in column "B" that are associated with the structure/description provided in column "A".

	A	B
1.	fasciculus gracilis	Gall
2.	fasciculus cuneatus	Burdach
3.	spinoolivary tract	Helweg / Bechterew
4.	mammillo-tegmental tract	Gudden
5.	anterior corticospinal tract	Turck
6.	dorsolateral fasciculus	Lissauer
7.	dorsal tegmental decussation	Meynert
8.	ventral tegmental decussation	Forel
9.	interstitio-spinal fasciculus	Muskens
10.	diagonal band	Broca
11.	stria medullaris	Piccolomini
12.	dorsal spinocerebellar tract	Flechsig
13.	ventral spinocerebellar tract	Gowers
14.	hook bundle (uncinate fasciculus of cerebellum)	Russell
15.	commissure of the lateral lemniscus	Probst
16.	dorsal supraoptic commissure	Meynert
17.	ventral supraoptic commissure	Gudden
18.	anterior hypothalamic commissure	Ganser
19.	sulcomarginal fasciculus	Lowenthal
20.	medial lemniscus	Reil
21.	septomarginal fasciculus	Bruce / Flechsig
22.	fasciculus interfascicularis (coma fasciculus)	Schultz / Hoche
23.	solitary (respiratory)tract	Krause / Gierke
24.	tectospinal tract	Held or Lowenthal
25.	rubrospinal tract	Monakow

26.	corticopontine tract	Fleschig
27.	fasciculus retroflexus	Meynert
28.	optic rediations	Gratiolet
29.	mammillo-thalamic tract	Vic d' Azyr
30.	lenticular fasciculus	Forel
31.	medial and lateral longitudinal striae	Lancisi
32.	dorsal trigeminothalamic tract	Wallenberg
33.	superior longitudinal fasciculus	Burdach
34.	fasciculus solitarius	Gierke
35.	habenulo-interpeduncular tract	Meynert
36.	frontopontine tracts	Arnold
37.	lateral spinothalamic tract	Edinger
38.	anterior spinothalamic tract	Dejerine
39.	septomarginal fasciculus	Flechsig
40.	tempropontine tracts	Turck / Arnold
41.	supramammillary decussation (commissure)	Forel
42.	stria terminalis	Foville / Tarin
43.	cingulum bundle (fornix periphericus)	Arnold
44.	tectopontine tract	Munzer
45.	tectoreticular tract	Pavlov
46.	dorsal longitudinal fasciculus	Schutze
47.	rubro-olivary-cerebellar-rubro triangle	Mollaret
48.	olivospinal tract	Helweg
49.	anterolateral fasciculus	Gower
50.	hippocampal commissure	Gudden
51.	psalterium (post part of fornix)	David
52.	dentate nuc to red nuc to olivary nuc to cerebellum	Mollaret
53.	spinal marginal tract	Spitzka
54.	geniculocalcarine projections in temporal lobe	Meyer / Archambault
55.	fasciculus proprius	Flechsig
56.	medial longitudinal fasciculus	Collier

57.	intracortical fibers in occipital lobe	Gennari
58.	mammillotegmental tract	Gudden
59.	prerubral fields	Forel
60.	decussating cochlear fibers forming lateral lemniscus	Held
61.	sulcomarginal fasciculus	Marie
62.	triangular fasciculus	Phillipe-Gombault
63.	olivocochlear bundle	Rasmussen
64.	decussation of superior cerebellar peduncle	Stilling
65.	limbic system circuit involved in memory	Papez
66.	inter-lenticular commissure in floor of III ventricle	Meynert
67.	myelinated fibers terminating in medial cervical ventral horn involved in head movements	Held

Neuroanatomy - Synonyms and Alternative Terms

Directions: Indicate the name of the individual(s) in column "B" that are associated with the structure/description provided in column "A".

	A	B
1.	ventral posterior medial nucleus (thalamus)	semilunar nucleus
2.	lateral cuneate nucleus	external, accessory
3.	hippocampal commissure	fornix transversus
4.	pars triangularis	Broca's cape
5.	petrotympanic fissure	Glasserian fissure
6.	gracile tubercle	clava
7.	nucleus gracilis	clavate nucleus
8.	nucleus cuneatus	triangular nucleus
9.	mammillary bodies	corpora candicans
10.	superior orbital fissure	sphenoidal fissure
11.	ciliary ganglion	lenticular or ophthalmic
12.	pterygopalatine ganglion	sphenopalatine ganglion
13.	facial canal	aqueduct of Fallopius
14.	hypoglossal canal	anterior condylar foramen
15.	superior colliculus	testes
16.	inferior colliculus	nates
17.	trochlear nerve	pathetic nerve
18.	pneumogastric nerve	vagus nerve
19.	supracallosal gyrus	induseum griseum
20.	insular gyri	gyri operti
21.	mammillary bodies	bulbs of the fornix
22.	lamina terminalis	lamina cinerea
23.	cavum septum pellucidum	Sylvian or 5th ventricle
24.	trigeminal nerve	trifacial nerve

25.	arachnoid granulations	Pacchionian bodies
26.	tentorial notch	Pacchionian foramen
27.	intracerebral periarterial subarachnoid space	Virchow-Robbins space
28.	tectospinal tract	ventral longitudinal fascic.
29.	facial colliculus	eminentia teres
30.	solitary tract	trineural tract
31.	fasciculus retroflexus	Meynert's commissure
32.	Meyer's loop	Archambault's loop
33.	ligament suspends incus from roof of middle ear	Arnold's ligament
34.	stria medullaris in 4th ventricle	Bergmann's cords
35.	radial fibers of ciliary muscle	Bowman's muscle
36.	helicotrema	Breschet or Scarpa hiatus
37.	diploic veins	Breschet's veins
38.	median raphe of the corpus callosum	Chaussier's line
39.	perilymph	liquor of Cotunnius
40.	glia forming glia-pial membrane	marginal glia of Held
41.	endoneurium	Henle's sheath
42.	cerebral aqueduct	iter
43.	vagal trigone	Arnold's area
44.	tuberculum cinereum	trigeminal tubercle
45.	pterygoid canal	Vidian canal
46.	subcallosal gyrus	Zuckerkandl's convolution
47.	striopallidal axons	Wilson's pencils
48.	auditory (pharyngotympanic) tube	Eustachian canal
49.	primary auditory cortex	Heschl's convolution
50.	motor language area	Broca's area
51.	sensory language area	Wernicke's area
52.	primary visual cortex	calcarine cortex
53.	olfactory radiations of Zuckerkandl	diagonal band of Broca
54.	limbic lobe	fornicate lobe
55.	anterior perforated space	olfactory tubercle

56.	midline invaginated part of telencephalon	telencephalon impar
57.	half-moon shaped fissure capping posterolateral extent of calcarine fissure	lunate fissure
58.	amygdala	archistriatum
59.	fusiform gyrus	occipitotemporal gyrus
60.	substantia innominate	substriatal gray
61.	superior longitudinal fasciculus	arcuate fasciculus
62.	superior fronto-occipital fasciculus	subcallosal fasciculus
63.	sleep arteries of the brain	carotid arteries
64.	inner layer of external capsule	theca lenticularis

Microscopic Anatomy - Skin

Directions: Indicate the name of the individual(s) in column "B" that are associated with the structure/description provided in column "A".

	A	B
1.	tactile nerve endings found in the cornium	Meissner
2.	tactile nerve endings in the deep layer of the epidermis	Merkel
3.	the areolar glands of the breast	Montgomery
4.	low threshold very rapidly adapting sensory receptors	Pacini
5.	tactile sensory nerve endings of the corium	Ruffini
6.	topical lines representing orientation of collagen fibers	Langer
7.	putative thermoreceptors	Krause
8.	spherical receptor with less thick capsule and less coiled nerve process than Krause end bulb	Golgi-Mazzoni
9.	cells in basal layer of epidermis containing catecholamines	Merkel-Ranvier

Microscopic Anatomy - Musculoskeletal

Directions: Indicate the name of the individual(s) in column "B" that are associated with the structure/description provided in column "A".

	A	B
1.	resorptive recesses in bone occupied by osteoclasts	Howship
2.	inner or osteogenic layer of the periosteum	Ollier
3.	connective tissue fibers between periosteum and bone	Sharpey
4.	canals within bone	Volkman
5.	opening in osteum of bone	Havers
6.	storage granules of endothelial cells lining blood vessels	Wiebel-Palade
7.	muscle fiber elevation at neuromuscular junction	Doyere
8.	neuromuscular junction	Rouget

Microscopic Anatomy - Respiratory

Directions: Indicate the name of the individual(s) in column "B" that are associated with the structure/description provided in column "A".

	A	B
1.	interalveolar pores	Kohn

Microscopic Anatomy - Cardiovascular

Directions: Indicate the name of the individual(s) in column "B" that are associated with the structure/description provided in column "A".

	A	B
1.	specialized conducting muscle cells of the heart	Purkinje

Microscopic Anatomy - Gastrointestinal

Directions: Indicate the name of the individual(s) in column "B" that are associated with the structure/description provided in column "A".

	A	B
1.	splenic cords or red pulp cords	Billroth
2.	digestive glands of the duodenum	Brunner
3.	the phagocytic cells of the liver sinusoids	Kupffer
4.	straight, tubular intestinal glands	Lieberkuhn
5.	eosinophilic granule-containing cells of the small intestine	Paneth
6.	aggregations of lymphoid tissue in the small intestine	Peyer
7.	inner layer of cells lining acinis of pancreas	Bernard
8.	mesenchymal cells within muscle layers of the gastrointestinal tract	Cajal
9.	splenic lymph corpuscles	Malpighian
10.	perisinusoidal spaces in liver	Disse
11.	liver acinus	Rappaport

Microscopic Anatomy - Urinary

Directions: Indicate the name of the individual(s) in column "B" that are associated with the structure/description provided in column "A".

	A	B
1.	renal collecting ducts	Bellini
2.	renal capsule	Gerota
3.	urethral glands	Littre
4.	paraurethral glands	Skene
5.	perirenal fascia	Colles
6.	inter-ureteric ridge	Mercier
7.	retro-renal fascia	Zuckerkandl
8.	renal columns	Bertin
9.	glomerular capsule	Bowman
10.	sphincter urethrae	Guthrie
11.	urinary trigone	Lieutaud
12.	urethral lacunae	Morgagni
13.	"U" shaped part of the nephron	Henle
14.	watershead between anterior and posterior branches of the renal arteries	Brodel
15.	convoluted tubule of the kidney	Ferrein
16.	rete testes	Haller
17.	renal pyramid	Malphigi

Microscopic Anatomy - Reproductive

Directions: Indicate the name of the individual(s) in column "B" that are associated with the structure/description provided in column "A".

	A	B
1.	testosterone producing cells of the testes	Leydig
2.	cells supporting the testicular epithelium	Sertoli
3.	straight tubules of testes	Haller
4.	coiled part of efferent ductules of testes	Haller

Microscopic Anatomy - Nervous System

Directions: Indicate the name of the individual(s) in column "B" that are associated with the structure/description provided in column "A".

	A	B
1.	largest cell of the cerebellar cortex	Purkinje
2.	myelin producing cell in the PNS	Schwann
3.	rough endoplasmic reticulum	Nissl
4.	flattened membranous cisternae in nerve cell body	Golgi
5.	small bundles of unmyelinated fibers in a peripheral nerve	Remak
6.	large pyramidal cells of the motor cortex	Betz
7.	junctions between adjacent Schwann cells	Ranvier
8.	incisures in peripheral myelinated nerve fibers	Schmidt-Lantermann
9.	radial glia	Bergmann
10.	feathered glial cells	Fananas
11.	endoneurium	Key – Retzius
12.	large pyramidal cells in calcarine cortex	Meynert
13.	longitudinal Schwann cells in regrowing axons	Bunger
14.	inhibitory interneurons in spinal cord	Renshaw
15.	ADH and oxytocin bodies in posterior pituitary	Herring
16.	small horizontal cells in the cerebral cortex	Cajal
17.	small rounded multipolar cells on cerebral cortex	Martinotti
18.	visual cortex laminae III	Bailey and Bonin / Cajal
19.	visual cortex "solitary" cells	Meynert
20.	visual cortex "star pyramids)	Bailey and Bonin

Microscopic Anatomy - Eye and Ear

Directions: Indicate the name of the individual(s) in column "B" that are associated with the structure/description provided in column "A".

	A	B
1.	apocrine glands of the eyelid	Moll
2.	sebaceous glands of the eye lashes	Zeis
3.	tarsal glands of the eyelid	Meibomian
4.	scleral venous sinus	Schlemm
5.	membrane between corneal epithelium and stroma	Descemet / Demour
6.	membrane between corneal endothelium and stroma	Bowman
7.	vestibular membrane of the cochlea	Reissner
8.	structure containing auditory receptor cells	Corti
9.	spiral canal between inner and outer pillar cells	Corti
10.	columnar cells on floor of cochlear duct	Claudius
11.	optic disc (blind spot)	Mariotte

Microscopic Anatomy - Endocrine

Directions: Indicate the name of the individual(s) in column "B" that are associated with the structure/description provided in column "A".

	A	B
1.	insulin-producing, cells in the pancreas	Langerhans
2.	inner layer of pancreatic acinus	Barnard

Developmental Anatomy - Embryology

Directions: Indicate the name of the individual(s) in column "B" that are associated with the structure/description provided in column "A".

	A	B
1.	sinus venosus	Cuvier
2.	ovarian follicle	Graaf
3.	paramesonephric duct	Muller
4.	cartilage of the 2nd branchial arch	Reichert
5.	embryonic mesonephric duct	Wolff
6.	ductus venosus	Arantius
7.	remnants of Wolffian duct in paroophoron	Kobelt
8.	embryonic urethral glands	Skene
9.	ductus arteriosus	Botallo
10.	longitudinal duct of the epoophoron	Gartner
11.	epoophoron	Rosenmuller
12.	cartilage of the 1st branchial arch	Meckel
13.	fissure between subarachnoid space and middle ear	Hyrtl
14.	neurenteric canal or blastoporic canal	Kovalevsky
15.	foramen ovale	Botallo
16.	ostium secundum	Born
17.	primitive node (knot)	Hensen
18.	ectodermal source of anterior pituitary gland	Rathke
19.	ganglion cells on CN XII	Froriep
20.	terminal junction of the cardinal veins	Cuvier
21.	oral ectodermal pouch forming anterior pituitary	Rathke
22.	diverticulum of the ileum	Meckel
23.	gelatinous connective tissue of the umbilical cord	Wharton
24.	remnant of Wolffian body at posterior side of the testicle	Giralde

25.	canal in the fetal sphenoid bone	Landzert
26.	mesonephric duct	Leydig
27.	para-umbilical veins	Sappey
28.	remnant of the mesonephric duct	Morgagni
29.	ovarian follicle	DeGraff

Section Three - Alphabetical Listing of Individuals

Abemathy, John (1764-1831)

Achilles (~1180BC)

Adamkiewicz, Albert W (1850-1921)

Addison, Christopher (1869-1951)

Aeby, Christopher Theodore (1835-1885)

Alber,

Alcock, Benjamin (1801–?)

Ammon (Egyptian deity)

Amussat, Jean Z (1796-1856)

Andersch, Carlos Samuel (1732-1777)

Arantius, Giulio Cesare (1530 1589)

Archambault, LaSalle (1879-1940)

Arnold, Friedrich (1803-1890)

Aschoff, Karl Albert Ludwig (1866-1942)

Aselli, Gaspare (1581-1625)

Auerbach, Leopold (1828-1897)

Ayala, G (1878-1943)

Bachmann, Jean George (1877-1959)

Baillarger, John Gabriel Francois (1809-1890)

Bailey, Percival (1892-1973)

Ball, Charles Bent (Sir) (1851-1916)

Barnard, F A

Berrettini, P C

Bartholin, Caspar (1655-1738)

Bauhin, Gaspard (1560-1624)

Baumgarten, Paul Clemens (1848-1928)

Baxter, Donald E

Bechterew, Vladimir Mikhailovich (1857-1927)

Beclard, Pierre Augustin (1785-1825)

Bell, Charles (1774–1842)

Bellini, Lorenzo (1643-1704)

Bergmann, G H (1781-1861)

Bernard, C (1813-1878)

Berry, James (1860-1946)

Bertin, Exupère Joseph (1712-1781)

Betz, Volodymyr Oleksiyovych (1834-1894)

Bichat, Marie-Francois Xavier (1711-1802)

Bigelow, Henry Jacob (1818-1890)

Bilroth, Christian Albert Theodor (1829-1924)

Blacius, Gerhard (1627-1682)

Blandin, Philipe Fredich (1798-1849)

Blumenbach, Johann (1752-1840)

Bochdalek, Vincent Alexander (1801-1833)

Bogros, Jean-Annet (1786-1823)

von Bonin, Gerhardt (1890-1979)

Bonnet, Amedea (1809-1858)

Born,

Botalli, Leonardo (1530-1587)

Bottcher, Arthur (1831-1889)

Bowman, William (1816-1892)

Boyer, Baron Alexis (1757-1833)

Bourgery, Marc-Jean (1797-1849)

Breschet, Gilbert (1784-1845)

Brissaud, Edouard (1852-1909)

Broca, Pierre Paul (1824-1880)

Brodel, Max (1870-1941)

Brodie, Benjamin (1783-1862)

Brodie, Benjamin C (1783-1862)

Brodmann, Korbinian (1868-1918)

Bruce, Alexander (1854-1911)

Bruch, Karl Wilhelm Ludwig (1819-1884)

Brucke, Ernst Wilhelm (1819-1892)

Brunner, Johann Konrad (1653-1727)

Buck, Gurdon (1807-1877)

Budge, Ludwig Julius (1811-1888)

Bunger, Otto (1858-1905)

Burdach, Karl F (1776-1847)

Cajal, Santiago-Ramon (1852-1934)

Calleja, Julian (1836-1930)

Calot, Jean Fransois (1861-1944)

Camper, Peter (1722-1789)

Casserio, Guilio (1552-1616)

Chassaignac, Edouard P (1804-1879)

Chaussier, Francois (1746-1828)

Chievitz, Johan H (1850-1901)

Chopart, Francois (1743-1795)

Ciaccio, Guiseppi Vincenzo (1824-1901)

Civinini, Philipo (1805-1844)

Clado, Spiro (1856-1905)

Clarke, Jacob Augustus Lockhart (1817-1888)

Claudius, Fredrich (1822-1869)

Cleland, John (1835-1925)

Cloquet, Jules Germain, (1790-1883)

Colles, Abraham (1773-1843)

Collier, James S (1870-1935)

Cooper, Astley Paston (1768-1841)

Corti, Alfonso (1822-1888)

Cotunnius, Domenico (1736-1782)

Cowper, William (1666-1709)

Crampton, Phillip (1777-1858)

Cruveillhier, Jean (1791-1874)

Cuvier, Georges Leopold (1769-1832)

Darkschewitsch, Liverij Osipovich (1858-1925)

Darwin, Charles Robert (1809-1882)

Daubenton, Louis (1716-1799)

David,

Deiter, Otto Freidrich Karl (1834-1863)

Dejerine, Joseph Jules (1839-1917)

Demour, Pierre (1702-1795)

Denonvilliers, Charles-Pierre (1808-1872)

Descemet, Jean (1732-1810)

Disse, Joseph (1852-1912)

Dogiel, Alexander Stanislavovich, (1852-1922)

Dorello, Primo (1872-1963)

Douglas, James (1675-1742)

Doyere, Louis (1811-1863)

Drummond, David (1852-1932)

Duncan, James M (1826-1890)

Dupre, Marsel (1886-1971)

Dupytren, Baron Guillaume (1777-1835)

Duverney, Joseph G (1648-1730)

Eberstaller, O

von Ebner, Victor (1842-1925)

Ecker, Alexander (1766-1829)

Edinger, Ludwig (1855-1918)

Eglis,

Ehrenritter, Johann (?-1790)

Erb, Wilhelm Heinrich (1840-1921)

Eustachi, Bartolomeo (1500-1510 to1574)

Exner, Sigmund (1846-1926)

Fallopi, Gabriele (1523-1562)

Fananas, Louis Ramon y Cajal (1857-1917)

Ferrein, Antoine (1693-1769)

Fielding, George H (1801-1871)

Flack, Martin William (1882-1931)

Flechsig, Paul Emil (1847-1929)

Fleischmann, Fredrich Ludwig (1806–1886)

Flood, Valentine (1800-1847)

Foltz, Jean Charles Eugène (1822–1876)

Fontana, Felice (1730-1805)

Forel, Auguste-Henri (1848-1931)

Foville, Achille-Louis (1799-1878)

Frankenhauser (1822-1894)

Fritsch, Gustav Theodor (1838-1927)

Frohse, Fritz (1871-1916)

Froriep, August (1849-1917)

Fuse,

Galeati, Domenico (1686-1755)

Galen, Claudius or (Clarissimus) (130-200)

Gall, Franz Joseph (1758-1828)

Gallaudet,

Ganser, Siegbert J M (1853-1931)

Garter, Hermann Treschow (1785-1827)

Gasser, Johann Laurentius (1723-1765)

Gennari, Francesco (1750-1797)

Gerdy, Pierre Nicolas (1797-1896)

Gerlach, Joseph (1820-1896)

Gerota, Dimitri (1867-1939)

Giacomini, Carlo (1840-1898)

Gierke, Hans Paul Bernhard (1847-1886)

Gillette, Eugene P (1836-1886)

Gimbernat, Antoni (1734-1816)

Giralde, Joachim Albino Cardosa Casado (1808-1875)

Glassenman,

Glasser, Johan Heinrich (1629-1675)

Glisson, Francis (1597-1677)

Golgi, Camillo (1844-1926)

Gombault, Francois Alexis Albert (1844-1904)

Gowers, William Richard (1845-1915)

Graaf, Regnier de (1641-1673)

Gratiolet, Louis-Pierre (1815-1865)

Gruber, Wenzel Leopold (1814-1890)

Grynfeltt, Joseph Casimit (1840-1930)

Gudden, Bernhard (1824-1886)

Guerin, Alphonse F M (1816-1895)

Guthre,

Guyon, Felix (1831-1920)

Guyon, Georges Charles (1876-1961)

Halban,

Haller, Albrecht (1708-1771)

Hannover, Adolph (1814-1894)

Hartmann, Robert (1831-1893)

Harvey,

Hasner, Joseph Ritter von Artha (1819-1892)

Hassler,

Havers, Clopton (1657-1702)

Heister, Lorenz (1683-1758)

Held, Hans (1866-1942)

Helvetius, (Schweitzer) Johann Friedrich (1630 – 1709)

Helweg, Hans K S (1847-1901)

Henle, Friedrich Gustav (1809-1885)

Hensen, Victor (1835-1924)

Hering, Ewald (1834-1918)
Herring, Percy (1872-1967)
Herophilus (330BC-260BC)
Heschl, Richard Ladislaus (1824-1881)
Hesselbach, Franz Kaspar (1759-1816)
Hey, William (1736-1819)
Highmore, Nathaniel (1613-1685)
Hilton, John (1805-1878)
His, Wilhelm (1863-1934)
Hoche, Alfred E (1865-1943)
Home, Everard (1756-1832)
Horner, Johan Fredrich (1831-1886)
Huguier, Pierre C (1804-1873)
Hortega, Pio (1882-1945)
Houston, John (1802-1845)
Howship, John (1781-1841)
Hueck, Alexander F (1802-18240
Humphrey, George M (1820-1896)
Hunter, John (1728-1793)
Huschke, Emil (1797-1858)
Huxley, Thomas Henry (1825-1895)
Hyrtl, Josef (1810-1894)
Jackson, Jabez N (1868-1935)
Jacobson, Ludwig L (1783-1843)
Jensen,
Keith, Arthur (1866-1955)
Kent, Albert F S (1863-1958)
Kerckering, Theodore (1640-1693)
Key, Ernst A H (1832-1901)
Kiesselbach, Wilhelm (1839-1902)
Kilian, Hermann F (1800-1863)

Killian, Gustav (1860-1921)

Klein, Edward E (1844-1925)

Kobelt, Georg Ludwig (1804-1857)

Koch, Walter (1880-?)

Kohlrausch, Otto L B (1811-1854)

Kohn, Hans Nathan (1866-1935)

Kolliker, Albert (1817-1905)

Kovalevsky, Alexander O (1840-1901)

Krause Friedrich Theodor (1797-1868)

Krause, Wilhelm (1833-1910)

Krukenberg, Adolph (1816-1877)

Kugel, M A

Kuntz, Albert (1879-1957)

Kupffer, Karl Wilhelm von (1829-1902)

Labbe', Charles (1851–1889)

Lallouetter, Pierre (1711-1792)

Lamier,

Lancisi, Giovanni Maria (1654-1720)

Landstrom, John (1869-1910)

Landzert, Theodor Bernhard (1833-1889)

Langenbeck, Bernhard Rudolph Konrad (1810-1829)

Langerhans, Paul (1847-4888)

Langer, Karl (1919-1887)

Langley, John Neport (1852-1925)

Lannelongue, Odilon M (1840-1911)

Lantermann, A J (1845-1898)

Lanz, Otto (1865-1935)

Larrey, Barron Dominique, (1766-1824)

Latarget, Andre (1877-1947)

Laumonier, Jean (1749-1818)

Lauth, Thomas (1785-1826)

Leber, Theodor Karl Gustav (1840-1917)

LeCat, Claude-Nicholas (1700-1780)

Lenhossek, Mihaly (1863-1937)

Lesshaft, Peter Frantsevich (1837-1909)

Leydig, Franz von (1821-1908)

Lieberkühn, Johann Nathaniel (1711-1756)

Lieutaud, Joseph (1703-1768)

Liliequist, Bengt (1923-2008)

Lisfranc, Jacques (1790-1847)

Lissauer, Heinrich (1861-1891)

Lister, Joseph (1827-1912)

Little, James Laurence (1836-1855)

Littre, Alexis (1654-1726)

Lockwood, Charles B (1856-1940)

Louis, Pierre Charles (1787-1872)

Lowenberg, Benjamin B (1836-1905)

Lowenthal, Wilhelm (1850-1894)

Lower, Richard (1631-1691)

Ludwig, Karl F W (1816-1895)

Luschka, Hubert (1820-1875)

Luys, Jules Bernard (1828-1897)

Mackenrodt, A K (1859-1927)

Maier, Rudolph Robert (1824-1888)

Maissiat, Jacques H. (1805-1878)

Malgaigne, Joseph Francois (1806-1865)

Malpighi, Marcello (1628-1694)

Mall, Franklin P (1862-1917)

Marie, Pierre (1853-1940)

Mayo, William T. (1861-1939)

Magendie, Francois (1783-1855)

Mauchart, Burkhard D (1696-1751)

Malpaghi, Marcello (1628-1694)

Mariotte, Edme (1620-1684)

Marshall, John (1818-1891)

Martinotti, Giovanni (1857-1928)

Mayer, August, Franz, Joseph (1787-1865)

Mazzoni, Vittorio (1880-1940)

McBurney, Charles (1845-1913)

McEwan, William (1848-1924)

Meckel, Johane Friedrich the elder (1724-1774)

Meckel, Johane Friedrich the younger (1781-1833)

Meibomian, Heinrich (1638–1700)

Meissner, Georg (1829-1903)

Mercier, Louis Auguste (1811-1882)

Merkel, Friedrich Sigmund (1845-1919)

Meyer, Adolf (1866-1950)

Meyer, George H (1815-1892)

Meynert, Theodor Herman (1833-1892)

Mohrenheim, Joseph J (1755-1799)

Moll, Jacob Anton (1832–1914)

Mollaret, Pierre (1898-1987)

Monakow, Konstantin (1853-1930)

Monro, Alexander (Secundus) (1733-1817)

Montgomery, William Fetherston (1797-1859)

Morgagni, Giovanni Battista (1682-1771)

Moskowitz, M

Morrison, James Rutherford (1853-1939)

Muir,

Muller, Heinrich (1820-1864)

Müller, Johannes Peter (1801-1858)

Munzer, Egmont (1865-1924)

Muskens, Louis Jacob Josef (1872-1937)

Nageotte, Jean (1866-1948)

Nelaton, Auguste (1807-1873)

Neubauer, Johann Ernst (1742-1777)

Nissl, Franz (1860-1919)

Nuck, Anton (1650-1692)

Nuhn, Anton (1815-1889)

Obersteiner, Heinrich (1847-1922)

Oddi, Ruggero (1864-1913)

Ollier, Louis Xavier Edouard Leopold (1825-1900)

Onuf-Onufrowicz, Bronislaw (1863-1928)

Pacchioni, Antonio (1665-1726)

Pacini, Filippo (1812-1883)

Palade, George (1912-1008)

Paladino, Giovanni (1540-1560)

Paneth, Joseph (1857-1890)

Pansch, Adolph (1841-1887)

Papez, James (1883-1958)

Parona, Francesco (1861-1910)

Passavant, Philip Gustav (1815-1893)

Pavlov, Ivan (1849-1936)

Pechlin, Carl Fredrich (1720-1796)

Pecquet, Jean (1622-1647)

Percheron, Gerard (1930-2011)

Perlia, Richard (1860-1931)

Petit, Antoine (1718-1794)

Petit, Jean Louis (1674-1750)

Peyer, Johann Conrad (1653-1712)

Philippe, Claudien (1866-1903)

Piccolomini, Arcangelo (1526-1605)

Pirogov, Nicolay (1810-1881)

Plender,

Poupart, Francois (1616-1708)

Probst, Moriz (1867-1923)

Purkinje, Johannes Evangelista (1787-1869)

Prussak, Alexander (1839-1897)

Quenu, Edward (1852-1933)

Ranvier, Louis Antoine (1835-1922)

Rappaport, Aron Moses (1904-1992)

Rasmussen, Grant (1834-1881)

Rathke, Martin Heinrich (1793-1860)

Rau, Johann J (1668-1719)

Redlich, Emil (1866-1930)

Reichert, Karl Bogislaus (1811-1883)

Reid, Robert William (1851-1939)

Reidel, Bernhard Moritz Carl (1846-1916)

Reil, Johann Christian (1759-1813)

Reisseisen, Franz D. (1773-1828)

Reissner, Ernst (1824-1878)

Rell, johann Christian (1759-1813)

Remak, Robert (1815-1865)

Renshaw, Birdsey (1908-1948)

Retzius, Anders Adolph (1796-1870)

Richter, August G (1742-1812)

Ridley, Humphrey (1653-1708)

Rinvus, Augustus Quirinus (1652-1723)

Riolan, Jean (1580-1657)

Rolando, Luigi (1773-1831)

Robbin, Charles-Philippe (1821-1885)

Roller, Fredrich William (1802-1878)

Rosenmuller, Johann Christian (1771-1820)

Rosenthal, Fredrich C (1780-1829)

Rotter, Josef (1857-1924)

Rouget, Charles Marie Benjamin (1824-1904)

Rouviere, Henri (1876-1952)

Ruffini, Angelo (1874-1929)

Russell, James Risien (1863-1939)

Ruysch, Fredrich (1638-1731)

Sandstrom, Ivar (?-1889)

Sappey, Marie P C (1810-1896)

Santorini, Giandomenico (1681-1737)

Savage, Henry (1810-1900)

Scarpa, Antonio (1746-1832)

Schlemm, Friedrich (1795-1858)

Schmidt, Henry D (1823-1888)

Schneider, Conrad Victor (1610/14-1680)

Schnitker,

Schuller, Karl H (1843-1907)

Schultz, Max Johann (1825-1874)

Schwalbe, Gustav Albert (1844-1916)

Schwann, Theodor (1810-1882)

Sebileau, Pierre (1860-1953)

Seiler, Carl, (1849-1905)

Sertoli, Enrico (1842-1910)

Servetus, Michael (1509-1553)

Sharpey, William (1802-1880)

Shrapnell, Henry J (1761-1841)

Sibson, Francis (1814-1876)

Skene, Alexander (1837-1900)

Soemmering, Samuel Thomas (1775-1830)

Spence, James (1812-1882)

Spieghel, Adriaan (1578-1625)

Spitzka, Edward C (1852-1914)

Staderini, Rutilio (1861–1942)

Steida, Ludwig (1837-1918)

Stenson, Niels (1638-1686)

Stilling, Benedict (1810-1879)

Struthers, John (1823-1899)

Suzanne, Jean G (1859- ?)

Sydney,

Sylvius, Francois de la Boe (1614-1672)

Symington, Johnson (1851-1924)

Tarin, Pierre (1725-1761)

Tawara, Sunao (1873-1952)

Tenon, Jacques-Rene (1724-1816)

Terson, Albert (1867-1935)

Thebesius, Adam Christian (1686-1732)

Theile, Fredrich, W (1801-1879)

Thompson, Allen (1809-1884)

Thompson, Henry (1820-1904)

Tiedmann, Fredrich (1781-1861)

Toldt, Carl (1840-1920)

Tome,

Torin,

Tortual Kaspar, (1802-1865)

Toynbee, Joseph (1815-1866)

Traub, Ludwig (1818-1876)

Treitz, (Wenzel) Vaclav (1819-1872)

Treves, Frederick (Sir) (1853-1923)

Trolard, Paulin (1842-1910)

Troltsch, Anton F (1829-1890)

Tsai,

Tulpius (Tulp) Nicolas (1593-1674)

Turck, Ludwig (1810-1868)

Turner, William (1832-1916)

Tyrrell, Fredrich (1797-1843)

Tyson, Edward (1651-1708)

Valentin, Gabriel Gustav (1810-1883)

Valsalva, Antonio (1666-1723)

Van Hoorne, Johannes (1621-1670)

Varolio Constantio (1543-1575)

Vater, Abraham (1684-1751)

Velpeau, Alfred Armand Louis Marie (1795-1867)

Vesalius, Andreas (1513-1564)

Vesling, Johann (1598-1649)

Vic d'Azyr, Filix (1748-1794)

Vidian, Guidi (1509-1569)

Vieussens, Raymond de (1641-1715)

Volkmann, Alfred Wilhelm (1800-1877)

Virchow, Rudolph (1821-1922)

Waldeyer, Heinrich Wilhelm Gottfried-Hartz (1836-1921)

Wallenberg, Adolf (1862-1949)

Walther, August F (1688-1746)

Weibel, Edward (1921-2019)

Weitbrecht, Josias (1702-1747)

Wepfer, Johan J (1620-1695)

Wernicke, Karl (1848-1905)

Westphal, Karl Friedrich Otto (1833-1890)

Wharton, Thomas (1610-1673)

Whitnall, Samuel Ernst (1876-1950)

Wilder,

Wilkie, David P D (1882-1938)

Willis, Thomas (1621-1675)

Wilson, James (1765-1821)

Wilson, Samuel Alexander Kinnier (1878-1937)

Winslow, Jacob Benignus (1669-1760)

Wrisberg, Heinrich August (1739-1808)

Wirsung, Johann Georg

Wolff, Kaspar Friedrich (1733-1794)

Wolfring, Emilj (1832-1906)

Wood,

Woolner, Thomas (1825-1892)

Worm, Ole (1588-1654)

Zaglas, John

Zeis, Edouard (1807–1868)

Zinn, Johann Gottfried (1727-1759)

Zuckerkandl, Emil (1849-1910)

About the Authors

Michael F. Nolan is professor of Basic Science Education at the Virginia Tech Carilion School of Medicine in Roanoke. He received his Physical Therapy training at Marquette University and his PhD in Human Anatomy from the Medical College of Wisconsin. Nolan spent the first 34 years of his career teaching gross anatomy and neuroanatomy to medical students and resident physicians at the University of South Florida. He has received more than 20 awards for excellence in teaching including the Master Teacher Award in 2014 from the International Association of Medical Science Educators and the John M. Thompson Outstanding Teacher Award in Neurosurgery in 2006. He has published over 40 peer-reviewed articles and book chapters as well as four textbooks in human gross anatomy and neuroanatomy.

John P. McNamara is the Director of Anatomy and Assistant Professor of Basic Science Education at the Virginia Tech Carilion School of Medicine in Roanoke. His doctoral training is in chiropractic from Life University (Marietta, GA) with undergraduate (Lock Haven University of Pennsylvania) and graduate (Shippensburg University of Pennsylvania) degrees. He is also ABD from Virginia Tech in Educational Leadership and Policy Studies. For nearly the past 30 years, McNamara has maintained a private practice in Salem, VA, and taught full-time anatomy and physiology, gross anatomy, neuroanatomy, and pathophysiology at the College of Health Sciences (Jefferson College) in Roanoke. From 2013 to 2017 he taught the gross anatomy course for the Doctor of Physical Therapy program at Radford University in Roanoke. He is licensed to practice as a Doctor of Chiropractic in both Virginia and Pennsylvania, and he is certified as an Emergency Medical Technician in Virginia.

Made in the USA
Coppell, TX
04 November 2026

11287990R00063